NJU SA 2022-2023

南京大学建筑与城市规划学院　建筑系教学年鉴
THE YEAR BOOK OF ARCHITECTURE DEPARTMENT TEACHING PROGRAM
SCHOOL OF ARCHITECTURE AND URBAN PLANNING NANJING UNIVERSITY
唐莲 李鑫 编　EDITORS : TANG LIAN, LI XIN
东南大学出版社・南京　SOUTHEAST UNIVERSITY PRESS, NANJING

建筑设计及其理论
Architectural Design and Theory

张 雷 教 授	Professor ZHANG Lei
冯金龙 教 授	Professor FENG Jinlong
吉国华 教 授	Professor JI Guohua
周 凌 教 授	Professor ZHOU Ling
傅 筱 教 授	Professor FU Xiao
王 铠 副研究员	Associate Researcher WANG Kai
钟华颖 副研究员	Associate Researcher ZHONG Huaying
吴佳维 副研究员	Associate Researcher WU Jiawei
梁宇舒 助理教授	Asistant Professor LIANG Yushu

城市设计及其理论
Urban Design and Theory

丁沃沃 教 授	Professor DING Wowo
鲁安东 教 授	Professor LU Andong
华晓宁 副教授	Associate Professor HUA Xiaoning
胡友培 副教授	Associate Professor HU Youpei
窦平平 副教授	Associate Professor DOU Pingping
刘 铨 副教授	Associate Professor LIU Quan
尹 航 讲 师	Lecturer YIN Hang
唐 莲 副教授	Associate Professor TANG Lian
曹竞文 博士后	Postdoctor CAO Jingwen
李 鑫 博士后	Postdoctor LI Xin

建筑历史与理论及历史建筑保护
Architectural History and Theory and Protection of Historic Buildings

赵 辰 教 授	Professor ZHAO Chen
王骏阳 教 授	Professor WANG Junyang
胡 恒 教 授	Professor HU Heng
冷 天 副教授	Associate Professor LENG Tian
史文娟 副研究员	Associate Researcher SHI Wenjuan
赵潇欣 博士后	Postdoctor ZHAO Xiaoxin
王洁琼 博士后	Postdoctor WANG Jieqiong

建筑技术科学
Building Technology Science

吴 蔚 副教授	Associate Professor WU Wei
郜 志 副教授	Associate Professor GAO Zhi
童滋雨 副教授	Associate Professor TONG Ziyu
梁卫辉 副教授	Associate Professor LIANG Weihui
施珊珊 副教授	Associate Professor SHI Shanshan
孟宪川 副研究员	Associate Researcher MENG Xianchuan

南京大学建筑与城市规划学院建筑系
Department of Architecture
School of Architecture and Urban Planning
Nanjing University
arch@nju.edu.cn http://arch.nju.edu.cn

教学纲要（2023版）
EDUCATIONAL PROGRAM

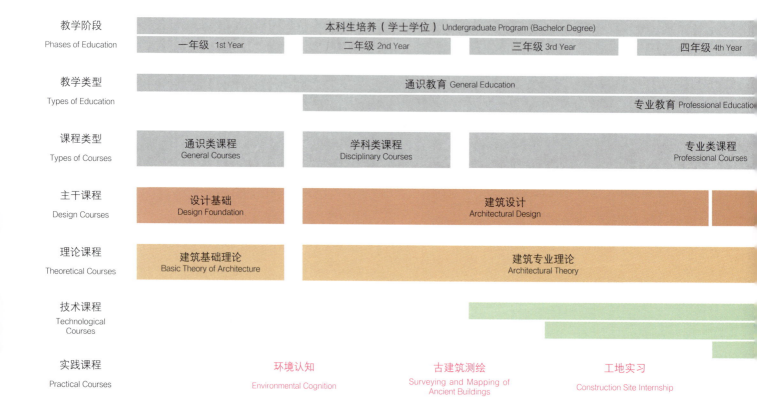

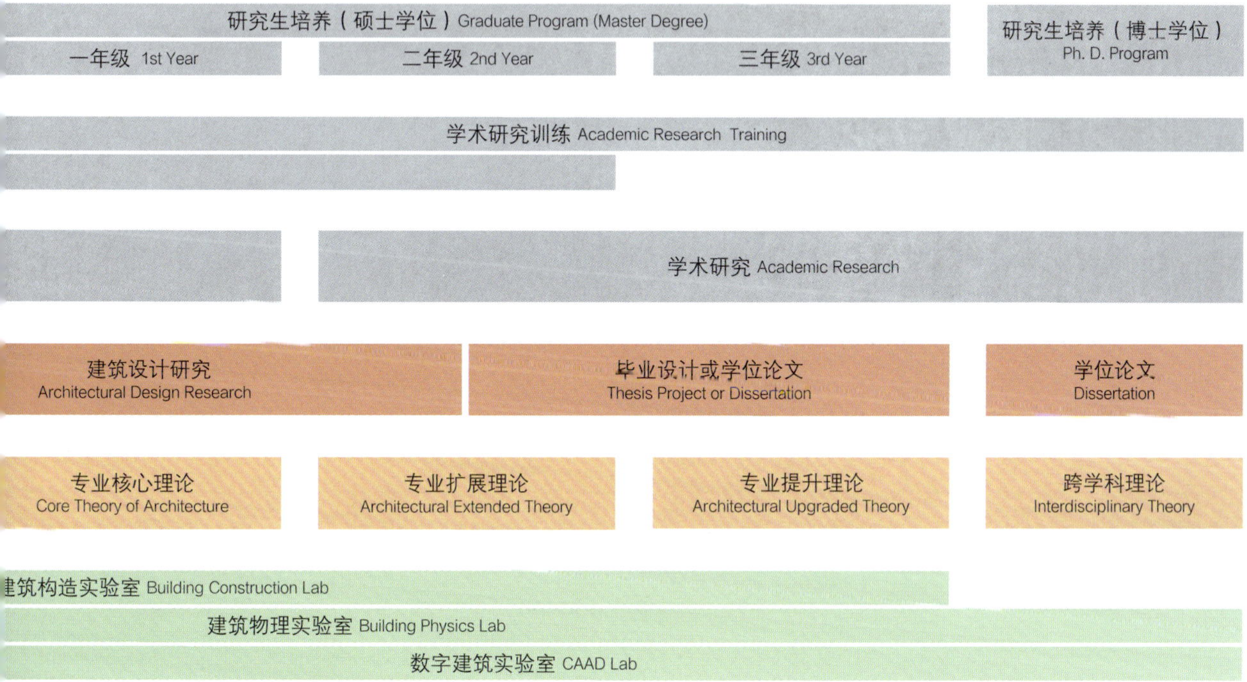

研究生培养（硕士学位）Graduate Program (Master Degree)			研究生培养（博士学位） Ph. D. Program
一年级 1st Year	二年级 2nd Year	三年级 3rd Year	

学术研究训练 Academic Research Training	

学术研究 Academic Research

建筑设计研究 Architectural Design Research	毕业设计或学位论文 Thesis Project or Dissertation	学位论文 Dissertation

专业核心理论 Core Theory of Architecture	专业扩展理论 Architectural Extended Theory	专业提升理论 Architectural Upgraded Theory	跨学科理论 Interdisciplinary Theory

建筑构造实验室 Building Construction Lab

建筑物理实验室 Building Physics Lab

数字建筑实验室 CAAD Lab

课程安排（2023版）
CURRICULUM OUTLINE

	本科一年级 Undergraduate Program 1st Year	本科二年级 Undergraduate Program 2nd Year	本科三年级 Undergraduate Program 3rd Year
设计课程 Design Courses	设计基础 Design Foundation	建筑设计（一） Architectural Design 1 建筑设计（二） Architectural Design 2 建筑与规划设计（一） Architecture and Planning Design 1 建筑与规划设计（二） Architecture and Planning Design 2	建筑设计（三） Architectural Design 3 建筑设计（四） Architectural Design 4 建筑设计（五） Architectural Design 5 建筑设计（六） Architectural Design 6
专业理论 Architectural Theory	建筑通史 General History of Architecture 建成环境导论与学科前沿 Introduction to the Built Environment and Frontiers of the Discipline	建筑设计基本原理 Basic Theory of Architectural Design 建筑与规划理论（一） Architecture and Planning Theory 1 建筑与规划理论（二） Architecture and Planning Theory 2	建筑设计基本原理 Basic Theory of Architectural Design 居住建筑设计与居住区规划原理 Theory of Housing Design and Residential Planning 城市更新规划* Urban Regeneration Planning
建筑技术 Architectural Technology		CAAD理论与实践（一） Theory and Practice of CAAD 1 建筑物理/建筑技术（二）：声光热 Building Physics / Architectural Technology 2: Sound, Light and Heat 建筑设备/建筑技术（三）：水电暖 Construction Equipment / Architectural Technology 3: Water, Electricity and Heating 建筑力学 Architectural Mechanics 建筑结构 Building Structure 建筑与规划技术（一） Architecture and Planning Technology 1 建筑与规划技术（二） Architecture and Planning Technology 2	建筑技术（一）：建构设计 Architectural Technology 1: Construction Design 建成环境科学* Built Environment Science
历史理论 History Theory		中国建筑史（古代） History of Chinese Architecture (Ancient)	外国建筑史（当代） History of Foreign Architecture (Modern) 中国建筑史（近现代） History of Chinese Architecture (Modern)
实践课程 Practical Courses			乡村振兴建设实践* Practice of Rural Revitalization Construction 城乡认知实习* Urban and Rural Cognitive Internship
大类通识类课程 Major General Courses	科学与艺术 Science and Art 现代工程与应用科学导论* Introduction to Modern Engineering and Applied Science 系统、决策与控制导论* Introduction to Systems, Decision-Making and Control 普通物理（力学）* General Physics (Mechanics) 大学化学* College Chemistry 管理学* Management 经济学原理* Principles of Economics 自动化导论* Introduction to Automation	社会学概论 Introduction to Sociology	
专业通识类课程 Professional General Courses	人居环境导论* Introduction to Human Settlement Environment	中外城市规划建设史* History of Urban Planning and Construction in China and Abroad 城市道路交通规划与设计* Traffic Planning and Design of Urban Roads 环境科学导论* Introduction to Environmental Science	

本科四年级 Undergraduate Program 4th Year	研究生一年级 Graduate Program 1st Year	研究生二、三年级 Graduate Program 2nd & 3rd Years
建筑设计（七）Architectural Design 7 建筑设计（八）Architectural Design 8 本科毕业设计 Graduation Project	建筑设计研究（一）Architectural Design Research 1 建筑设计研究（二）Architectural Design Research 2 研究生国际教学工作坊* Postgraduate International Design Studio	专业硕士毕业设计 Thesis Project
建筑设计行业知识与创新实践 Knowledge and Innovative Practice in the Architectural Design Industry 城市设计及其理论* Urban Design and Its Theory 设计研究与环境行为* Design Research and Environmental Behavior 景观规划设计及其理论* Landscape Planning Design and Its Theory	城市形态与设计方法论 Urban Morphology and Design Methology 建筑与城市研究基础 Foundations of Architecture and Urban Studies 现代建筑设计基础理论 Preliminaries in Modern Architectural Design 城市形态学* Urban Morphology 景观都市主义理论与方法* Theory and Method of Landscape Urbanism	
CAAD理论与实践（二）Theory and Practice of CAAD 2 建筑师业务基础知识 Introduction to Architects' Profession 建筑节能与绿色建筑 Building Energy Efficiency and Green Building 建成环境科学* Built Environment Science 建设工程项目管理* Construction Project Management	建筑体系整合 Building System Integration 建筑学中的技术人文主义 Technology of Humanism in Architecture 建筑环境学与设计 Architectural Enviromental Science and Studies GIS基础与应用* Concepts and Application of GIS 材料与建造* Materials and Construction 计算机辅助技术* Computer-Aided Technology 传热学与计算流体力学基础* Fundamentals of Heat Transfer and Computational Fluid Dynamics 数字建筑设计* Digital Building Design 绿色建筑技术* Green Building Technology	
	建筑理论研究 Studies of Architectural Theory 建筑史研究* Studies of the History of Architecture 中国木建构文化研究* Studies in Chinese Wooden Tectonic Culture	建筑设计实践 Architectural Design and Practice
工地实习* Construction Site Internship 古建筑测绘* Surveying and Mapping of Ancient Buildings		

课程说明：* 表示选修课程

本科建筑与规划实验班课程安排
EXPERIMENTAL UNDERGRADUATE PROGRAMME FOR ARCHITECTURE AND PLANNING

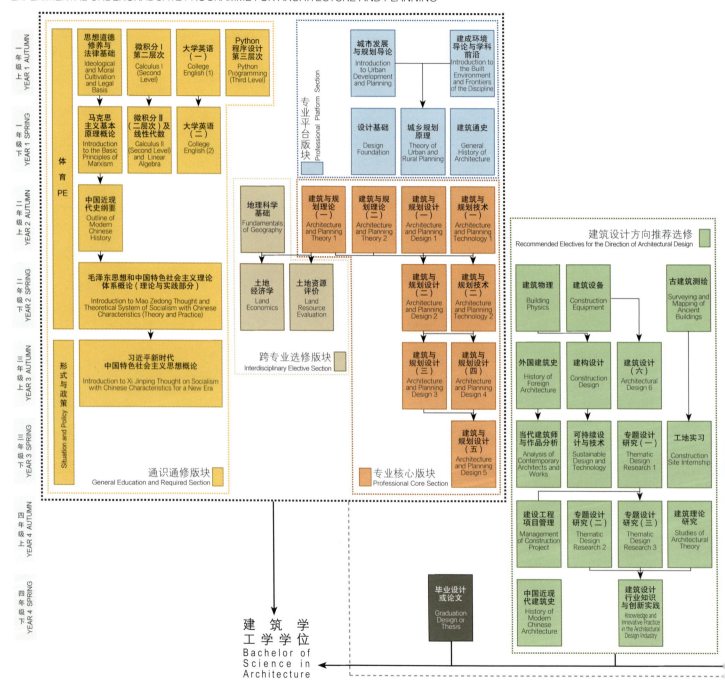

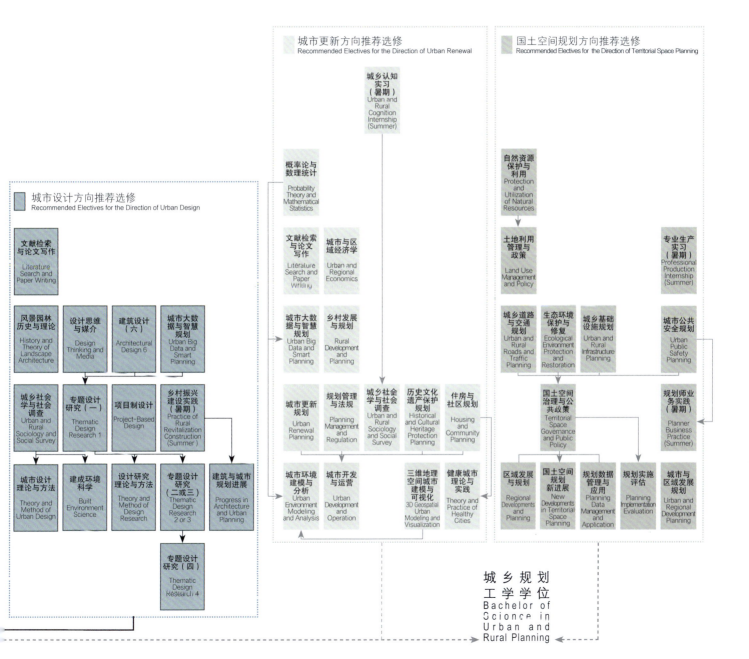

教学论文 TEACHING PAPERS

2
空间化的"闻过楼"——一项"建筑史方法"课程的作业
THE SPATIALIZED "TOWER OF ACCEPTING CRITICISM"—AN ASSIGNMENT FOR THE COURSE "METHODS OF ARCHITECTURAL HISTORY"

12
从古画到影像——明·吴彬《端阳》图的一次移情重构
FROM ANCIENT PAINTINGS TO IMAGES—AN EMPATHETIC RECONSTRUCTION OF WU BIN'S PAINTING *DUANYANG* (*DRAGON BOAT FESTIVAL*) IN THE MING DYNASTY

课程概览 COURSE OVERVIEW

22
建筑通史
GENERAL HISTORY OF ARCHITECTURE

24
建成环境导论与学科前沿
INTRODUCTION TO THE BUILT ENVIRONMENT AND FRONTIERS OF THE DISCIPLINE

26
设计基础
DESIGN FOUNDATION

32
建筑设计（一）：限定与尺度——独立居住空间设计
ARCHITECTURAL DESIGN 1: LIMITATION AND SCALE—INDEPENDENT LIVING SPACE DESIGN

36
建筑设计（二）：校园多功能快递中心设计
ARCHITECTURAL DESIGN 2: CAMPUS MULTI-FUNCTIONAL EXPRESS CENTER DESIGN

40
建筑与规划设计（二）：旧城中的新社区
ARCHITECTURE AND PLANNING DESIGN 2: NEW COMMUNITY IN OLD TOWN

44
建筑设计（三）：都市田园共享公寓
ARCHITECTURAL DESIGN 3: URBAN GARDEN SHARED APARTMENT

48
建筑设计（三）：基于算法的专家公寓设计
ARCHITECTURAL DESIGN 3: THE EXPERT APARTMENT DESIGN BASED ON ALGORITHM

52
建筑设计（三）：集装箱中学设计
ARCHITECTURAL DESIGN 3: THE DESIGN OF CONTAINER MIDDLE SCHOOL

56
建筑设计（四）：世界文学客厅
ARCHITECTURAL DESIGN 4: WORLD LITERATURE LIVING ROOM

60
建筑设计（五）：大学生健身中心改扩建设计
ARCHITECTURAL DESIGN 5: RECONSTRUCTION AND EXPANSION DESIGN OF COLLEGE STUDENT FITNESS CENTER

68
建筑设计（六）：社区文化艺术中心设计
ARCHITECTURAL DESIGN 6: DESIGN OF COMMUNITY CULTURE AND ART CENTER

76
建构设计研究：工地实习
CONSTRUCTION DESIGN RESEARCH: CONSTRUCTION SITE INTERNSHIP

82
建筑认知实践：古建筑测绘
ARCHITECTURAL COGNITIVE PRACTICE: SURVEYING AND MAPPING OF ANCIENT BUILDINGS

目 录

88
建筑设计（七）：高层办公楼设计
ARCHITECTURAL DESIGN 7: DESIGN OF HIGH-RISE OFFICE BUILDINGS

94
建筑设计（八）：城市设计
ARCHITECTURAL DESIGN 8: URBAN DESIGN

100
本科毕业设计
GRADUATION PROJECT

112
建筑设计研究（一）：基本设计
ARCHITECTURAL DESIGN RESEARCH 1: BASIC DESIGN

130
建筑设计研究（一）：概念设计
ARCHITECTURAL DESIGN RESEARCH 1: CONCEPTUAL DESIGN

142
建筑设计研究（二）：综合设计
ARCHITECTURAL DESIGN RESEARCH 2 : COMPREHENSIVE DESIGN

160
建筑设计研究（二）：城市设计
ARCHITECTURAL DESIGN RESEARCH 2 : URBAN DESIGN

172
研究生国际教学工作坊
POSTGRADUATE INTERNATIONAL DESIGN STUDIO

1—19 教学论文 TEACHING PAPERS

20—181 课程概览 COURSE OVERVIEW

183—194 建筑设计课程 ARCHITECTURAL DESIGN COURSES

195—197 建筑理论课程 ARCHITECTURAL THEORY COURSES

199—202 城市理论课程 URBAN THEORY COURSES

203—205 历史理论课程 HISTORY THEORY COURSES

207—211 建筑技术课程 ARCHITECTURAL TECHNOLOGY COURSES

213—215 认知实习 COGNITIVE INTERNSHIP

217—223 其他 MISCELLANEA

教学论文
TEACHING PAPERS

教学论文 TEACHING PAPERS

空间化的"闻过楼"——一项"建筑史方法"课程的作业
THE SPATIALIZED "TOWER OF ACCEPTING CRITICISM." — AN ASSIGNMENT FOR THE COURSE "METHODS OF ARCHITECTURAL HISTORY"

胡恒
HU Heng

1. 故事

明朝嘉靖年间,直隶常州府宜兴县有位"殷太史",官拜侍讲之职。其表兄弟名顾呆叟,乃顾恺之后裔。两人自小相熟,十分契密。呆叟生性恬淡,与友相处耿直忠言,广负贤名。殷太史好其敬若神明,爱同骨肉。呆叟科场不利,三十开外就生白须,决定干脆弃了举业,归隐山林。呆叟去城四十余里购置了茅舍薄田,打算终老于此。

呆叟自入山中,快活似神仙。众友屡番写信求其回转,呆叟坚拒。尤其殷太史思其人而不得,遂在临别谈心的一楼前自题一匾名"闻过楼"。

半年后的一日,县中差人上门,说官家签派呆叟入城管下年收粮的监兑之事。呆叟不愿回城,即取百金"行贿",脱了这个差事。"大放血"之后,呆叟竭力经营,半年后才恢复元气。正待好好打理自己的"世外桃源"之时,又逢盗匪来劫,将家中细软一扫而空,还狠似无意地留下几件东西。呆叟只当盗匪从其他地方劫掠所得,遗漏于此,就放在一旁不理。遭劫之后,呆叟穷困非常。城中旧友寄来书信安慰,但无人伸以援手。呆叟熬穷守困,好歹撑了下来。一日,又有差人来捉他,说是一伙盗匪被捕,招认说有些赃物落在顾姓某人家中。呆叟一看,正是前番那些无名之物。呆叟长叹曰,清福难享,更甚于富贵。就着家属收拾行李,起身回城。差人倒是和气,路上并未难为于他。

行到近城数里之外,许多车马停在道边。呆叟准备低头掩面过去,但几人走过来拽住,说其主人与呆叟有事相商。拉扯之际,几位大佬从附近一处村落赶来,要请呆叟过庄一叙。差人也不甚阻拦,呆叟遂应允。大家坐定,叙了寒温。呆叟将自己的遭遇说了一遍,众人出主意,让他先在庄园中住下,待这些朋友去县尊处沟通,帮他免去见官的尴尬。呆叟同意。

第二日,一位差役到访,正是前番通知其入城公干的那个人。他将本为"行贿"之用的百金还给呆叟,说该差事改签他人,这百金没用上,现在呆叟回城,他觉得还是还回为妙。又过了一日,那些劫掠过他的盗贼再闯进来,说是误劫高人名士,理当赔罪,就把原物送还。呆叟十分愁苦,认为赃物归还之事太过匪夷所思,一旦见官,必然有理说不清。呆叟正在胡思乱想之际,县尊突然到访,表示希望呆叟能够如之前那样,让大家可以朝夕请教,并且生计方面不用担忧,他自有安排。

呆叟乍惊乍喜之际,众友携酒而来为其压惊道贺。吃到半席,众人才将内中详情告知呆叟。自其入山之后的种种怪事,其实都是大家合力谋划所为,只为让呆叟享不了林泉之福。正巧新来的县尊怜才惜士,所以大家伙同县尊一起做局,逼其从山中回归城内。那个山庄也正是大家一起集资买给呆叟的,让他安心住在不城不郭之间,既方便大家继续交流思想,又能成就其高尚之心。

呆叟听后并未怪罪生气,服从大家安排。殷太史在其隔壁又置一间别业,将"闻过楼"匾额搬过来钉上,让呆叟朝夕相规,不时劝诫。

图1 《闻过楼》插图

图 2 [元] 王蒙《葛稚川移居图》

2. 山居、"整蛊"与"脱隐"

《闻过楼》是李渔《十二楼》中的最后一篇，其为半自传半小说，与《三与楼》相似。自传部分取自李渔早年归隐山林的一段经历，小说部分以此为导引，延伸成篇。全文共三回。第一回前半截，李渔现身说法山居之乐，随后进入正题，小说男主如李渔一般义无反顾地走上归隐之路。这一回的内容虚实相杂，大多取材自李渔自己。第二、三回为小说主体，是相当少见的"整蛊"喜剧——《十二楼》中仅此一例。一帮"损友"合谋将男主逼出山林，回归集体（图1）。李渔将"一段林泉佳话、麈尾清谈"与一出荒诞的整蛊闹剧创造性地拼接在一起，立意新颖，笑点频频。

李渔"入山"的经历相当丰富。1644 年，正值明清交替，李渔 33 岁，住在金华。乱世之中，兵火延烧全国，李渔时常去乡下山中避难。比较长的"入山"时间是 1646 年（顺治三年），清兵到金华，李渔携家带口避祸到兰溪下李村故居，一直住到 1650 年世道平静下来才搬家去杭州。该故居在离村子一里之外的伊山脚下。李渔归隐后将其改造成著名的"伊山别业"。《闻过楼》第一回引用的《山斋十便》绝句就出自其《伊园十便》。李渔另外还有《伊园杂咏》《伊园十二宜》等组诗，可见其在山中的日子相当惬意。《闻过楼》里，呆叟入山前后的种种心态无疑都是李渔自己的写照。区别在于，李渔是为了避祸，而呆叟则是厌倦俗世，不堪市井纷扰。

小说二、三回的整蛊游戏全为杜撰。男主归山之后，一帮损友做局，让呆叟在一年半之内连遭三桩横祸，不仅倾家荡产，还惹上盗匪官司，穷途末路、颠沛流离之下还要被押解回城见官，说不定还要吃上牢饭。这个整蛊之局布置得精巧周密、环环相扣，甚至拉上一县之尊参与，动用了公门的权力，颇有天罗地网的架势，纵然呆叟人品超卓、才情无双也难逃此劫。

李渔没有经历过这样的整蛊闹剧，但显然希望整蛊背后的事情发生在自己身上。做官的朋友视之为挚友，采其诤言；有钱的朋友为其买宅置产，安排生活；在不村不郭之间安居，享受两边的好处，进退自如。当然，现实是残酷的，这些梦想只能在小说故事里成真。

3. 画史中的《闻过楼》

故事的一个内核是可图像化。山居、招隐在古代画史中是经典主题，相关画作十分丰富。文中的引经据典——比如桃源、剡溪、希夷、南郭、陶潜、林逋、晋文公、介子推——都是常见的绘画题材，甚至"半村半郭之间""避秦之地"也都有许多相近的画作留存。换而言之，去掉整蛊闹剧，《闻过楼》的故事几乎可以用现成的图画来解说一遍。李渔将呆叟身份设定为顾恺之后人，也有图在文外的意思。

首先是入山。故事开篇有一首李渔自写的"入山避难诗"。诗中所言与元代王蒙的《葛稚川移居图》正相对应（图2）。画中描绘的是东晋道士葛洪为避战乱携家带口移居罗浮山修道的故事。葛洪身着鹤氅，手摇羽扇，牵着一头驴代步。身后跟着妻儿老小，牵牛携仆往山中行进（图3）。这一景象大概与李渔 1646 年为避清兵举家入伊山无异。

图 3 [元] 王蒙《葛稚川移居图》局部　　　　　　　　　　　　　　　　　　　　　　　　图 4 [明] 董其昌《荆溪招隐图》

文人的避世归隐是一种普遍理想。李渔在文中但凡言及于此就兴致盎然。看得出他对这种生活的着迷。这类画作中以明代董其昌的《荆溪招隐图》与元代赵孟頫的《谢幼舆丘壑图》为代表。呆叟的归隐之地亦在荆溪——"就在荆溪之南，去城四十余里，结了几间茅屋，买了几亩薄田，自为终老之计"。董其昌的《荆溪招隐图》是 1613 年送给朋友宜兴名流吴澈如的一幅水墨长卷（图 4），意思是两人虽然都在朝为官，但均有招去归隐的意愿。画中的山石、树木、溪水、茅舍的布置安排较为复杂，没有董其昌其他山水画中的清爽斩截的构图风范，但真实性反而有所增强。《闻过楼》里的故事发生在明嘉靖的宜兴，也即，呆叟的隐居之地就在董其昌的这幅画中。以董其昌及这幅画的名声，我们假设李渔对《闻过楼》故事的空间设定就来自此，也不无可能。

故事中，呆叟在荆溪隐居，并未真的避世到人迹罕至的深山。他只是想离城市生活尽可能远，不受其干扰。隐居期间，他没有像那些文人所想的那样过着眠云漱石、坐忘参禅的精神生活，而是正儿八经地农圃种植，营运资生，做一个踏踏实实的"为农之儒者"。仇英有许多"山居"题材的画作。他的《辋川十景图》中有一个庄园环境与文中相近。一个小小的院落，屋前有黑犬，篱笆院内有石台，屋后有两个干草堆（图 5）。似乎呆叟就在这里和家人过着平凡且平静的日子，就此终老。《莲溪渔隐图》中远有群山，近有绿水，水旁有田，田边为宅，一如呆叟"几间茅屋，几亩薄田"的愿景（图 6）。

4. 可图像化的与不可图像化的

所以，《闻过楼》的故事可以分为两个部分：可图像化的和不可图像化的。前者几乎串起半部画史，后者即是整盅游戏。后者之所以难以被图像化，不在于作者为追求戏剧性而虚构过度或普遍性不足，而在于其中隐含着呆叟的心理世界。这个世界并非如文中描述的那样恬淡寡营、傲骨铮铮，也有难以示人的一面。这一面被刻意掩藏起来，因为一旦被揭开，整个故事就会崩塌。

尽管掩藏较深，但这"难以示人"的一面在整盅游戏中仍留有蛛丝马迹——游戏中存在一些漏洞。似乎李渔在此设下一个局，等待读者进来破解。

第一处"漏洞"在入山前。呆叟科举不利准备"招隐"。"告了衣巾，把一切时文讲章与镂管穴孔的笔砚，尽皆烧毁，只留下农圃种植之书，与营运资生之具。"入山后，过了第一次行贿之劫，呆叟刚刚恢复元气，就想着"构书屋于住宅之旁，蓄蹇驴于黄犊之外"。既有"书屋"，就必有笔墨纸砚、图书卷轴；既有"蹇驴"，就必有远游、访友计划。可见入山前的那番激烈举动，只是向别人表明自己开始农家新身份的坚定态度。而时过境迁，稍稍经历一点折腾就暴露本心，打算重启"书屋""远游"的文人生活。前后立场不一，呆叟显出"表演型人格"。

第二处"漏洞"是从山上回来后。在撤退之前，他的朋友建议他住在"城村之间"，在那里不仅可以远离喧嚣，还可以和朋友相聚。呆叟不屑一顾，认为住在城村之间，而不是偏远的山区，可能会比在城中有更多的活动，所以断然拒绝。然而，当他别无选择时，他认为这也是一个好地方——"若非陶处士之新居，定是林山人之别业"，"朴素之中又带精雅，恰好是个儒者为农的住处"，"称谢不遑"。呆叟态度从"断然拒绝"到"称谢不遑"的彻底转变，表明呆叟引以为傲的"理智豁达"的品质也只是一句空话。

《闻过楼》的两套结构浮出水面。可图像化的部分撑起了故事的外皮。一系列由文字衍生出来的古画，包裹起这半部自传，各种典故、传说纷至沓来。不可图像化的部分潜藏在外皮之内。它是呆叟的心理世界，其对此讳莫如深。三处"漏洞"掀开了呆叟内里一角，但在炫目外皮的遮蔽下难以被察觉。如果不是整盅游戏所迫，它们会如春风化雨，消失在"荆溪"之中。虽然隐藏，但是我们不能忽略这些漏洞，

图 5 [明] 仇英《辋川十景图》局部　　　　　　　　　　　　　　　图 6 [明] 仇英《莲溪渔隐图》局部

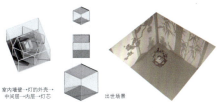
室内墙壁→灯的外壳→
中间层→内层→灯芯

图 7 《芯宅》设计概念

图 8 《芯宅》各层框架设计意向图

图 9 外框、内框不同角度的叠合

因为它们是探查呆叟(以及李渔自己)心理世界的线索——正如李渔在文末所说,"殊不知作者原有深心"。

5 两间"闻过楼":呆叟的心理世界

如何以空间体的形式再现呆叟的心理世界,探寻李渔的"原有深心",成为我"建筑史方法"课教学中的一项课题。2018 年与 2021 年出现的两份作业,从不同角度给出解答。

2018 年的作业名为"呆叟之心·2018"。作业的切入点是呆叟内心之光点燃 / 熄灭效果二重性(也即矛盾性)的可视化。我称之为《芯宅》。《芯宅》由两层外框与一支蜡烛组成。蜡烛位于最内层,灯芯即呆叟的内心。靠近蜡烛的中间层外框构造复杂,代表呆叟对外在世界的认知。这一外框分上下两层。下层是一圈市井图像,材质不透光;上层是镂空的花纹,有山林溪洞、花鸟鱼虫等图案。靠外的外框为透明材质,代表殷太史的倾心维护(图 7~图 9)。

蜡烛点燃时,光线穿过中层框架,其镂空的上部分的图案被投影出去。这些图案成为巨大的影子,穿过殷太史之框,映在更外层的房间的墙壁上。如果我们缓缓转动中层框架,墙壁上的影子也会随之动起来——树枝在动,鱼在游,鸟在飞。这个图景就是旁人看到的呆叟,"生在衣冠阀阅之乡,常带些山林隐逸之气"(图 10)。

蜡烛熄灭时,从外面来看,里面的状况什么都看不到。中层框架把灯芯挡得严严实实,只能看到框架下部的各种市井图像。这些图像与外部世界是一样的。而从里面看,中层框架就像四面高墙,将灯芯围得死死的。虽然墙的上部有诸多花纹,但毫无灵动自由之意,反倒像是监狱围墙顶上的铁丝网。

灯火明灭的两种状态产生的效果反差与呆叟的内心变化十分吻合。灯芯燃起,就是呆叟走出房门,与外在世界交流的时候。他想要表现的"恬澹(淡)寡营,志在山野",只是大家乐于看到和需要的。灯芯熄灭时,就是呆叟关上大门,回归本心之时。闭上双眼,他看到的是真实的内心——科举失利,让自己心如死灰,世界如此精彩,但已经没有勇气再踏足其间,只能一走了之,避迹深山,至少不用再整日戴着面具,伪装恬淡于人前。不然长此以往,恐怕要精神分裂(图 11)。

2021 年的作业名为"呆叟之声·2021"。这是一个空间/声音装置,原型为蝈蝈罐。它是呆叟的心之宅,也是声之屋,简称《过宅》(装置上刻有一个"过"字)(图 12、图 13)。

《过宅》的外壳分两圈。内圈与基座为一体,外圈连着"屋顶"。内、外圈都有两种开口:全开口和格栅式开口。当外圈旋转时,整个《过宅》会出现三种形态模式:第一种是内外开口叠合,《过宅》大门洞开;第二种是外圈的实墙挡住内圈的开口,《过宅》全封闭;第三种是外圈的格栅部分挡住内圈的开口,《过宅》半

图 10 蜡烛点燃时的《芯宅》俯瞰、外层投影

图 11 蜡烛熄灭时的《芯宅》俯瞰、外观

图 12 《过宅》上部的"过"字

图 13 《过宅》全景

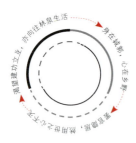  

图 14 《过宅》概念分析

封半闭。第一种全开型对应的是呆叟科举求仕，以及与朋友们车马往来的城市生活；第二种全封闭型对应的是呆叟携家入山，归隐山林；第三种半封半闭型对应的是半村半郭之地（图 14）。

内外圈的错动、叠合关系，意味着呆叟的现实状况（固定的内圈），是由心理世界（滑动的外圈）来决定的。比如，在城市的时候，两个开口应该叠在一起，表示开门迎接来访友人。如果呆叟心情不好，准备烧笔摔砚，或者是向友人大谈违心之言的时候，外圈的开口就会滑到内圈的格栅处，表示内（心）外（表）错位的"表演"开始了。在山林里的时候，本来应该是外圈开口对着内圈的实墙，表示与世隔绝，只守着耕作纺织的最底线生产。但是一旦手有余钱，生计宽松，就想着再构书房，养一头代步蹇驴，恢复文人的浪游生活。这时，外圈的开口就滑开，与内圈的开口叠合，表示呆叟准备再度出发，游山玩水去也（图 15）。

《过宅》将呆叟的内心世界表现得十分全面，各种暧昧之处都能转化成外圈的轻巧滑移。除此之外，它还是一个声音装置。《过宅》内放一只蝈蝈，不同的空间状态，其声音也会不同：或兴奋长鸣，意指出门赴考场，求取功名；或时低时高，意指坦腹山林，逍遥自在；或尖叫紧张，意指遭遇盗贼，噤若寒蝉；或时停时叫，意指居于半村半郭，相互闻过。心由声达，闻"过"即闻"蝈"（图 16）。

6. 结语

《芯宅》与《过宅》都实现了对呆叟心理世界的探寻。我们似可理解了李渔以《闻过楼》终结《十二楼》的用意。在原有体例之外，它还是半篇画史，更是一部心理小说。故事在此完成了两次跳跃：第一次是外向的图像化；第二次是内向的精神分析化。这一点更具意义——自传升级为自我分析（尽管是无意识），意味着《闻过楼》离开了话本传奇的旧有范畴，跨进全新的叙事领域。

分析、展现李渔这两次跳跃，就是本文的目标。外部目标较易达到。李渔以顾恺之为引子，以典故为铺垫，使得该图像化工作具有天然的系统性。我们只需按字索图、按图索即可完成。其中，董其昌、仇英贡献了实景参考。内部目标相对麻烦。一则，心理世界本就难以图示化，而所言皆虚、所行皆伪的呆叟的内心更难窥测；二则，那些作为线索的逻辑漏洞不能视为叙事不严密所致，只有联系李渔自身经历才能理解其合理性；三则，以空间体再现心理世界，欠缺先例参照。虽然困难多多，《芯宅》与《过宅》仍然达成了目标。它们以心之"灯"与心之"声"为切入点，建构起两座心之"宅"，展现出呆叟心理世界的复杂、矛盾性，回应了李渔的期待。

图 15 《过宅》的几种开口方式

图 16 《过宅》的声音装置

1. Story

During the reign of Emperor Jiajing of the Ming Dynasty, there was a "Grand Historian Yin" in Yixing County, Changzhou Prefecture, Zhili Province, who held an official post of Imperial Tutor. He and his cousin, Gu Daisou, a descendant of Gu Kaizhi, had been intimately acquainted with each other since childhood. Daisou was a calm and straight talker and was well known for his honesty, to whom Grand Historian Yin gave much love and esteem. After suffering setbacks in imperial examinations, Daisou was white-bearded in his 30s, and then he decided to abandon his attempt in examinations and resort to seclusion in the mountains. He bought a thatched cottage and some small plots of farmland more than forty li away from the town, where he planned to stay to the end of his life.

Since then, Daisou had lived a leisurely life in the mountains. Several letters from his friends asking him back were firmly refused. In particular, Grand Historian Yin missed him so much that he inscribed a plaque named "The Tower of Accepting Criticism" on the gate top of the tower where they had a parting talk.

One day half a year later, a runner came from the county saying that the goverment had dispatched Daisou to supervise the grain collection of the subsequent year in the town. Daisou was unwilling to go back, so he "bribed" the runner with a hundred taels of silver to free him from this errand. The loss of so much money forced him to work hard for half a year before recuperation. While he was about to take good care of his home, bandits came and ransacked his house, with only a few things left unintentionally. Daisou thought they were taken from other places by the bandits and left here, just putting them aside. This robbery put Daisou in dire economic straits. Letters of comfort came from old friends in the town, but no help was offered. Daisou managed to survive the poverty. One day, another runner came to catch him, saying that a gang of bandits arrested confessed that some stolen things were left at a Gu's home. On closer inspection, Daisou found they were those things from nowhere. With a deep sigh, he said that having a carefree life was even harder than having wealth. He told his family to pack up and then he set off for the town. The runner was kind and did not make difficulties for him on their way back to the town.

Many horse-drawn carriages were parked by the roadside a few li from the town. When Daisou was about to pass by the road with his head down and features hidden, several men held him and said that their master had a matter to discuss with him. Just then, some well-respected old men rushed out from a nearby village and invited Daisou to have a talk with them in a villa. Seeing that the runner did not stop them, Daisou agreed. They sat down and greeted each other. Then Daisou recounted his story. Other men advised him to stay here for the time being, and they would go to talk over the matter with the magistrate of the county, so that he did not need to meet the official. Daisou agreed.

The next day, there was a visit from the runner who had sent word to Daisou to supervise grain collection. He came here to return the "bribe" to Daisou, explaining that someone had taken over the work and the silver was not needed anymore, and it would be better to return it now that Daisou was back in town. Another day, those bandits burst in again, saying that they had robbed a man of great virtue by mistake and should make amends, and then they handed all things back to Daisou. Feeling quite puzzled, Daisou thought it was too weird, and he could not explain it clearly once met the official. When he was thinking about this, the magistrate showed up unexpectedly, expressing that he hoped to get along with Daisou like before and that he had made arrangements for his livelihood and he did not have to worry.

In the midst of surprise and joy, Daisou was greeted with his friends and wine. Halfway through the feast, Daisou was told of the details. All the strange things that had happened after his seclusion were planned, just to deprive him of pastoral pleasures. It happened that the new magistrate cherished talented people, so they intentionally set Daisou up to force him back from the mountains. The villa between the town and the village was bought with the money collected for Daisou, so that he could be free to stay here and they could continue to exchange ideas with him without compromising his noble heart.

Instead of blaming them, Daisou accepted this arrangement. Grand Historian Yin bought another house in the next door and moved the plaque "The Tower of Accepting Criticism" here, so that he could always be admonished by Daisou.

2. Life in the mountains, "prank" and "return from seclusion"

Tower of Accepting Criticism, the last volume of *Twelve Towers* written by Li Yu, consists of an autobiographical section and a novelistic section, and is similar to *Sanyu Tower*. Materials of the autobiographical section are drawn from Li's experience of retreating to the mountains in his early years, and the novelistic section is a continuation of the autobiographical section. Among the three chapters of *Tower of Accepting Criticism*, Li describes the joy of living in the mountains based on his own experience in the first half of chapter one, leading to the subject that the hero of the novel embarked on the road of seclusion without hesitation. This chapter combines virtuality with reality, most of which is predicated on Li's own life. Chapters two and three, the main body of the novel, are a rare "prank" comedy, the only one in *Twelve Towers*. A group of "mischievous friends" conspired to force the hero back from the mountains into their group (Figure 1). Li artfully combines "a much-told seclusion story" with a ridiculous prank, which is original and amusing.

Li had a quite rich experience of retreating to the mountains. In 1644 when the dynasty was alternating from Ming to Qing, Li, at the age of 33, was living in Jinhua. In the turbulent days, the country was scorched by war, and Li often went to the rural mountains to escape the fighting. He has stayed in the mountains for a relatively long time since 1646 (in the third year of the reign of Emperor Shunzhi) when the Qing army came to Jinhua. He, with his family, fled to his former home in Xiali Village, Lanxi and stayed there until 1650 when the war temporarily ceased and they moved to Hangzhou. The home at the foot of Yishan Mountain, a li from Xiali Village, was transformed into the famous "Yishan Villa" by Li after his seclusion. "Ten Conveniences of the Mountain Hut", the jueju (a poem of four lines) quoted in Chapter one of *Tower of Accepting Criticism* came from his "Ten Convenience of the Yi Garden". Besides, Li also wrote such suite poems as "Miscellaneous Poems of the Yi Garden" and "Twelve Proprieties of the Yi Garden", which showed how pleasant he was in the mountains. In *Tower of Accepting Criticism*, the changes in Daisou's state of mind before and after seclusion were undoubtedly the portrayal of Li himself, except for the difference that Li retreated to the mountains to escape the war and Daisou was tired of the secular world and wanted to stay off the busy life.

The prank in Chapters two and three of the novel is completely fabricated. After Daisou, the hero of the novel, retreated to the mountains, a group of his friends set him up, and finally he, in a year and a half, lost everything, got into troubles with bandits, and was escorted to the town when in desperation and might even be put into prison. This prank was so well-planned that it even involved the magistrate of the county with the power of the officialdom, even if Daisou of excellent character and ability could not escape from it.

Li did not have, and yet obviously wished to have, such an experience. Friends in government saw him as an intimate friend and accepted his straightforward criticism; wealthy friends bought a dwelling and arranged life for him; settling down between the town and the village enabled him to enjoy the convenience of both the town and the village. Of course, reality is cruel, and these dreams can only come true in the novel.

3. *Tower of Accepting Criticism* in the painting history

One core element of the story is that it can be picturized. Life in the mountains and seclusion are classic themes in the ancient painting history, and there are also abundant related paintings. Allusions in the texts, including the Land of Peach Blossoms, the Shanxi River, Xiyi, Nanguo, Tao Qian, Lin Bu, Duke Wen of Jin and Jie Zitui, are common painting subjects, and there are even works similar to the subjects "being between the town and the village" and "a place away from war". In other words, the story of *Tower of Accepting Criticism*, if not included the prank, can almost be interpreted with the existing paintings. By setting Daisou as a descendent of Gu Kaizhi, Li also intends to express the relationship between paintings and texts.

The first part is a retreat to the mountains. The story begins with a poem "Seek Shelter in the Mountains" written by Li, whose content corresponds to the painting *Ge Zhichuan Yiju Tu* (*Ge Zhichuan Moving Residence*) created by Wang Meng in the Yuan Dynasty (Figure 2). This painting depicts the story of Ge Hong, a Taoist in the Eastern Jin Dynasty, moving to Luofu Mountains with his family and living a monastic life there to escape war. Ge is dressed in a fur cloak with crane pattern, with a feather fan in one hand and a donkey held by the other, followed by his family, heading for the mountains (Figure 3). This painting depicts a situation similar to Li's moving to Yishan Mountains with his family in 1646 to escape the Qing army.

Seclusion is a universal ideal of literati. Li shows great interest every time he writes about this topic in his work, which shows his fascination with this life. Representative paintings of this kind are *Jingxi Zhaoyin Tu* (*Secluded Dwelling in Jingxi*) of Dong Qichang in the Ming Dynasty and *Xie Youyu Qiuhe Tu* (*The Mind Landscape of Xie Youyu*) of Zhao Mengfu in the Yuan Dynasty. Daisou was also secluded in Jingxi—"built several cottages and bought a few fields in the south of Jingxi, more than forty li away from the town, where he planned to spend his remaining years till death". *Jingxi Zhaoyin Tu* (*Secluded Dwelling in Jingxi*) is a long scroll of ink painting that Dong presented to his friend Wu Cheru, a celebrity in Yixing, in 1613, expressing their desire to seclude themselves from the imperial court where they were working (Figure 4). Unlike the simple and neat style of his other landscape paintings, the rocks, trees, brook and cottages in this painting are arranged in a more complex manner, but are more realistic. The story of *Tower of Accepting Criticism* took place in Yixing during the reign of Emperor Jiajing of the Ming Dynasty. That is, Daisou's secluded dwelling is in this painting. By the reputation of Dong Qichang and this painting, perhaps we can assume that Li's setting the story of *Tower of Accepting Criticism* was inspired by this painting.

In the story, Daisou lived in seclusion in Jingxi, instead of really retreating to the remote mountains with few inhabitants. He just wanted to get as far away from the town as possible to avoid its disturbances. As a recluse, he did not immerse himself in a spiritual life as literati thought, but lived steadily by engaging in serious farming. Among many works themed "life in the mountains" made by Qiu Ying, a garden in his *Wangchuan Shijing Tu* (*Ten Views in Wangchuan*) is similar to the environment described in the story. In a small courtyard, there is a black dog in front of the house, a stone table inside the fence yard, and two haystacks behind the house (Figure 5). It seems that Daisou lived an ordinary and peaceful life with his family to the end of his life. In *Lianxi Yuyin Tu* (*Seclusion as Fishermen by Lianxi Brook*), there are mountains in the distance and green water nearby, by the water is a field and by the field stands a house, just as Daisou's desire for "several cottages and fields" (Figure 6).

4. Picturization and non-picturization

Therefore, the story of *Tower of Accepting Criticism* can be divided into two parts: a part that can be picturized, and a part that cannot be picturized. The former almost strings together half of the painting history, and the latter, the prank game, is difficult to picturize because it implies Daisou's mental world rather than due to the author's excessive and less universal story-making for dramatic effect. The world is not as unworldly and upright as described in the story, and it also has its unmentionable secrets. These secrets are deliberately hidden, because once they are revealed, the whole story will collapse.

Deeply hidden as they are, there are still signs of these "unmentionable" secrets in the prank game, namely some loopholes in the game. Li seems to set a puzzle here for readers to solve.

The first "loophole" is before retreating to the mountains. Daisou was ready for "reclusion" after failing imperial examinations. "After receiving the honorary title from the imperial court, he burned all essays, carved writing brushes and inkstones, leaving only farming books and tools." After retreating to the mountains, Daisou met with his first catastrophe of bribery. When he just got back on his feet, he thought of "building a book house next to his dwelling, and raising a lame donkey next to cattle". With a "book house", he must prepare writing brushes, ink sticks, paper, inkstones, books and scrolls; with a "lame donkey", he must also have a plan to travel and visit friends. This means that what he did before retreating to the mountains was just to show to others his determination to start a new life as a farmer. However, as time passed, he quickly exposed his heart and wanted to resume his literati life with a "book house" and "travel". This inconsistency shows the "histrionic personality" of Daisou.

The second "loophole" is after returning from the mountains. Before retreating, his friends advised him to live "between the town and village", where he could not only get away from the hustle and bustle, but also get together with his friends. Daisou was dismissive, thinking that he, living between the town and village rather than the remote mountains, might have more activities with others than in the town, so he refused this absolutely. As he was running out of options, however, he thought this was also a nice place — "This is either a new residence of Tao Qian, or another house of a recluse in the mountains", "austere and delicate, this is such a good place for a literatus to live a farming life", "thank you without hesitation". The complete change of Daisou's attitude from "refusing this absolutely" to "thank you without hesitation" indicates that the quality of being the "sensible and open-minded" that Daisou was proud of is also just an empty phrase.

There appear the two structures of *Tower of Accepting Criticism*. The picturized part is the framework of the story. A series of ancient paintings depicted by words make up this autobiographical section, in which there are a number of allusions and tales. The non-picturized part, namely Daisou's mental world that is hard to describe, is hidden in the framework. The three "loopholes" hazily present Daisou's mental world, but are inconspicuous under the complicated framework. They may disappear unconsciously into "Jingxi", without the prank game. Hidden though they are, we cannot ignore them, since they are clues to the mental world of Daisou (and Li himself) — just as what Li said at the end of the text, "it turned out that that was exactly what the author originally intended".

5. Two towers of accepting criticism: Daisou's mental world

How to reproduce Daisou's mental world and explore what Li "originally intended" in the form of a space has become a topic in my teaching of the course "Methodology of Architectural History". Two assignments in 2018 and 2021 have answered this question from different perspectives.

The assignment in 2018 titled "The Heart of Daisou · 2018". The trigger with this assignment is the visualization of the dual igniting/extinguishing effect of Daisou's inner light (contradiction), which I call Xin Zhai (The Core House). Xin Zhai is composed of two frames and a candle. Among them, the candle is located in the innermost layer, with the wick as Daisou's heart. The middle-level frame near the candle is complex, representing Daisou's cognition of the outside world. It consists of two layers, of which the lower layer is covered with marketplace images and made of opaque materials, and the upper layer is engraved with hollow patterns including mountains, streams, flowers, birds, fish and insects. The outermost frame is made of transparent materials, representing Grand Historian Yin's admiration and consideration (Figure 7 ~ Figure 9).

When the candle is lit, the light passes through the middle-level frame, the hollow patterns in the upper layer are projected into huge shadows, and the shadows pass through the outermost frame, namely the frame of Grand Historian Yin, and reflected on the wall of the outer room. If we slowly turn out the middle-level frame, the shadows on the wall will also move—branches are moving, fish are swimming and birds are flying. This scene is the image of Daisou in the eyes of others, "a person born in a family of merit, with a recluse temperament" (Figure 10).

When the candle is extinguished, nothing inside can be seen from the outside. As the wick is completely blocked by the middle-level frame, only various marketplace images in the lower layer of the frame can be seen. These images are the same as the outside world. From the inside, the wick is totally blocked by the middle-level frame, like by four walls. Although many patterns can be seen on the wall, they are like the barbed wire on the top of the prison walls, showing no flexibility and freedom.

The contrast between igniting and extinguishing effects is very consistent with the changes in Daisou's inner world. When the wick is ignited, it represents that Daisou walks out of the room and communicates with the outside world. The "indifference to fame and wealth, and desire for seclusion" that he wants to express is just what others are willing to see and need. When the wick is extinguished, it means that Daisou closes the gate and returns to his true heart. Closing his eyes, he sees in his true heart that he feels utterly hopeless after failing imperial examinations, and he has no fortitude to embrace the world although it is so wonderful. He has no choice but hide himself in the remote mountains, where at least he does not have to disguise himself as being indifferent to fame and wealth. Otherwise, he may be mentally dissociated in a long run (Figure 11).

The assignment in 2021 is "The Sound of Daisou · 2021". This is a space/sound device, prototyped by a grasshopper jar. It is Daisou's house of heart as well as his house of sound, which is referred to as Guo Zhai (a Chinese character pronounced as "guo" is engraved on the upper part of the device) (Figure 12 and Figure 13).

The shell of Guo Zhai is divided into two circles, of which the inner circle and the chassis are combined into one, and the outer circle is connected to the "roof". Both circles have two kinds of openings: full opening and grille-type opening. When the outer circle rotates, there will be three morphological patterns in the whole Guo Zhai. The first is the openings of the inner and outer circles are superimposed, and the gate of Guo Zhai is wide open. The second is the solid wall of the outer circle blocks the opening of the inner circle, and Guo Zhai is totally closed. The third is the grille of the outer circle blocks the opening of the inner circle, and Guo Zhai is semi-closed. The first pattern represents that Daisou attends the imperial examinations for an official and lives a city life with his friends. The second pattern represents that Daisou retreats to the mountains with his family. The third pattern represents that Daisou settles down between the town and the village (Figure 14).

The staggering and superimposition of inner and outer circles suggest that Daisou's realistic condition (the fixed inner circle) is determined by his mental world (the sliding outer circle). For example, when Daisou is in the town, the two openings should be superimposed, indicating that he opens the door to welcome his friends. When Daisou is in a bad mood and is ready to burn his writing brushes and break his inkstones, or when he says things contrary to his convictions to his friends, the opening of the outer circle will slide to the grille of the inner circle, representing that Daisou starts to hide his heart. When Daisou is in the mountains, the opening of the outer circle should have been opposite the solid wall of the inner circle to indicate that Daisou lives in seclusion and is engaged only in farming and textile. Yet, once he is well off, he thinks of building a book house and raising a lame donkey for riding, leading his wandering life as a literatus. At this time, the opening of the outer circle will be superimposed with that of the inner circle to indicate that Daisou plans to travel (Figure 15).

Guo Zhai has fully reflected Daisou's inner world, and all ambiguities can be expressed by the slides of the outer circle. Besides, it is also a sound device. With a grasshopper in Guo Zhai, its sound will be different in different spatial states: the long scream means that Daisou attends the imperial examination for an official; the rising and falling of the sound means that Daisou retreats to the mountains and lives a carefree life; the nervous scream means that Daisou is robbed by bandits and is afraid; the intermittent scream means that Daisou lives between the town and the village and is open to criticism. Daisou's inner life is expressed by the sound of the grasshopper. Being opening to criticism is reflected by listening to the grasshopper (here, "过", which means criticism, and "蝈", which means grasshopper, are homophonous in Chinese) (Figure 16).

6. Conclusion

Both Xin Zhai and Guo Zhai have successfully explored Daisou's mental world. We seem to understand Li's intention of ending *Twelve Towers* with *Tower of Accepting Criticism*. In addition to its original style, it is also half of the painting history, and even a psychological novel. The story has realized two leaps. The first is the outward picturization, and the second is the inward psychoanalysis. The latter is even more meaningful—the autobiography is upgraded to the self-analysis (even it is unconscious), meaning that *Tower of Accepting Criticism* develops from the original category of storytelling legend to a new narrative field.

Analyzing and revealing the two leaps is the objective of this paper. The external objective can be easily achieved. This paper begins with Gu Kaizhi and is built with the allusion, providing a natural system for the picturization of the story. We only need to understand the words in this paper and find paintings according to the text. We can refer to the works of Dong Qichang and Qiu Ying for the scenery. It is relatively troublesome to achieve the internal objective. First, the mental world can be hardly picturized, while the inner life of Daisou who always disguises himself with words and deeds contrary to his convictions can be more hardly seen. Second, the logical loopholes that are used as clues cannot be dismissed as the result of imprecise narrative, and only by combining Li's experience can we understand their rationality. Third, there is a lack of precedent reference to reproduce the mental world in the form of a space. Despite these challenges, Xin Zhai and Guo Zhai have still realized the objective. Triggered by the "light" of heart and the "sound" of heart, they built two "houses" of heart to display the complex and contradictory mental world of Daisou in response to Li's expectation.

教学论文 TEACHING PAPERS

从古画到影像——明·吴彬《端阳》图的一次移情重构
FROM ANCIENT PAINTINGS TO IMAGES — AN EMPATHETIC RECONSTRUCTION OF WU BIN'S PAINTING DUANYANG (DRAGON BOAT FESTIVAL) IN THE MING DYNASTY

郭烁　胡恒
GUO Shuo　HU Heng

1. 秦淮与端午

自古以来，秦淮河便是南京城最具代表性的空间符号之一。明代中后期，秦淮河畔的繁华达到一个高峰，尤其是城内的内秦淮一带（主要是南侧河段）聚集了无数文人骚客。这里白天熙熙攘攘，喧哗热闹，夜里更是橹声灯影，暗香浮动。秦淮河一年中最热闹的日子当属端午，这一天，城中百姓齐聚河岸，举办最具节日氛围的集体活动——赛龙舟。明《正德江宁县志》载："好事者买舟载酒戏游，谚云游舡，此俗近年最盛。"另外明末清初张岱的《陶庵梦忆》记载了竞渡的盛况："看西湖竞渡十二三次，己巳竞渡于秦淮，辛未竞渡于无锡，壬午竞渡于瓜州，于金山寺。"

时至今日，关于明代秦淮河上赛龙舟盛况的文字记录并不少见。更为难得的是，还有描绘此中情景的《端阳》一画流传于世，这幅画能让后人更为直观地感受到最繁华的城市空间（秦淮河）在节日氛围的高潮时刻（端午）的景象。

2. 古画《端阳》

《端阳》是一套名为《岁华纪胜图》的纪实册页组图中的一幅，由明代画家吴彬所绘（图1）。此组图以十二时令为主题，展现了太平盛世的繁华景象。虽然画作没有具体题词说明，但诸家研究者几乎都有共识，画中场景应该取自明代南京城——大约在1595—1612年间。除画面写实、手法精细外，此图册另有巧妙构思。十二幅画面分别以每月特殊节令设景，如果按照时间顺序依次展开，即可窥探明代南京城一年间的光阴流转、四季变迁。如进一步将每一幅画面联想成一节电影剧情，则不同节令的社会活动，不同地点的城市风貌，都在一幕幕"场景"中生动呈现。

画面长宽比约1∶2，视角呈轴测状，水体为中心，两岸左右上斜对作为配景，以桥相连。水上廊桥、水中沙渚、水面船帆共同构成了画面主体部分，桥上人物视线聚向水面，顺之可见一二龙舟隐于树叶摇曳之间。明代南京城进行赛龙舟的地方

图1 [明] 吴彬《端阳》

一般都在内外秦淮两处。外秦淮河段在大报恩寺及乌衣巷前的河道一直到长干桥及中华门附近。内秦淮河段则在青溪渡到贡院及夫子庙一段（图2）。《端阳》中最主要的"景点"是跨越河道的廊桥。这个廊桥设计得极其华丽。桥身较长，桥墩曲线向内弯，很是优雅。廊桥的中间是一座二层高的"塔楼"，内设有一个"佛龛"，高大的佛像和周围的一圈小佛像都清晰可见（图3）。这样精致奢华的廊桥不太可能出现在外秦淮河上——因为外秦淮河的桥基本上都有防御战事的功能，其形态都很朴素。所以，初步分析之下，画中场景应该就在现存的夫子庙及文德桥东端的某处。那座华丽的廊桥应该是当年河南岸的回光寺前的"利涉桥"——现在则是文正桥。《端阳》应为取自明末端午节内秦淮河畔赛龙舟的图景。

《端阳》上演的"秦淮与端午"大戏就此拉开帷幕，画面正中的一座小岛和人们乘船登上滩渚祭庙、观赛、赏景的场景，构成充满野趣的踏青场景；画面左下角桥头旁紧凑的商铺、忙碌的店家、买货的顾客构成热闹的市集买卖场景；左上方廊桥内，龙舟赛的观众凭栏助威，好奇的行人驻足张望；右上角山林茂密，近处沿岸的是茶楼，远处是村屋，氛围宁静，充满意趣；而画面的右下角，一位背猴男子似是急于赶上这场热闹，借机表演杂耍，大赚一笔。"午餐既竟，相率看龙舟"，商铺买卖、桥头观赛、小岛踏青、山林酒屋等闹静杂处的场景在眼前依次呈现。

3. 从图像到影像：《端阳·2023》视觉实验

短片《端阳·2023》是一次从历史研究出发，结合图像分析法和数字技术，将古画影像化表达的尝试。而完成的影像的重点在于回应古画中的"意蕴"。

根据上文分析，《端阳》画面空间上对应的原址地点位置较易确定，但是画中的"古意"却是有些隐晦。画面以水为中心、以桥为空间制高点的鸟瞰构图显然是为了应对水面活动赛龙舟的火热场景，而作为"主角"的龙舟在画中却极其不显眼——一条龙舟隐在画面左侧的一株柳树的树枝之间，一条龙舟藏在一座巨大的桥墩后方，只露出2个划手、3支红色船桨以及一小截船身。如果不留意搜寻，它们几乎不会被发现。桥上的观众、沙渚岸边的人群并不少，他们或探头，或扭颈，目光投射的地方却是空空的水面，气氛略显迷惑（图4）。这无疑是画家吴彬的独特设定，火热的场景还未登场，大家正在等待桥下冲出第一支龙舟的那一瞬间。这也许是真正身临其境的古人才理解的"赛龙舟"的趣味所在。

另一个隐而不显的"古意"是河两侧的人居状况。河下侧沿岸都是商铺与酒肆，河上一侧房屋则要讲究许多。一些屋舍后方是颇具规模的落落。它们在岸边伸出眺台，窗前坐着的都为女性，她们装扮得都很漂亮，摇着团扇闲聊观景（图5）。按照彼时的状况，河边有着眺台的院子都属于"旧院"（青楼）范畴，那些凭窗眺望的女子也是秦淮风月传说的一部分。当下的现实中，这些"古意"没有留下痕迹。除了秦淮河水依旧流淌之外，一切都已烟消云散。想要重现那个"古意"似乎并不容易。

图2 明代秦淮河赛龙舟路线粗略图

图3 《端阳》（局部）图中桥上的"塔楼"与"佛龛"

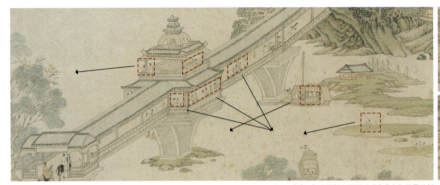  

图 4 《端阳》（局部）图面"观赛龙舟"场景分析　　图 5 《端阳》（局部）图中"旧院"空间

4. 剧本："探寻"

剧本拟照初见其貌、再观其形、后现其意的分析节奏，以赏画者的思绪为"主角"，以"探寻"为线索，沿着从古到今、从画内到画外的顺序展开剧情。全剧共三节：首先开篇表现端午节日的热闹场面，对应初见画面时的直观感受，着重渲染欢乐的气氛高潮；随后，走入画中，亲自"探寻"画中趣味；最后，在出画后回到"现实"——当下的秦淮河，于此"寻找"500年前的记忆痕迹。

4.1 第一节：龙舟戏秦淮

开篇的"主角"被热闹场景感染，不自觉化身为龙舟划手，与众人一起摇动双桨，挥汗如雨，激情澎湃。画面场景均采用数字建模技术进行模拟，视角在疾驰的各个龙舟附近及岸边鼓掌的观众身上不断切换。这部分的镜头需要多角度变化、速度感、特写及大空间俯瞰交叉切换（图6）。

该场景还有一个要点，那就是背景空间的特殊性。内秦淮这一河段的空间形态相当丰富，且带有强烈的南京特色。除了我们在《端阳》中看到的廊桥外，还有河房露台、岸边、沙渚、支流木桥、河边小船。贡院前的大片平整"广场"也可以作为"观景台"容纳大量人群聚集。河边的回光寺或许还有二层楼阁、塔之类的高视点观景处。细数之下，此处有近10种观看赛龙舟的空间位置。这些高低错落、犬

牙交错的观景点（市集、青楼、佛寺、贡院、廊桥、沙渚、画舫、小舟、短桥）一起融合进"龙舟戏秦淮"的狂野"开篇"之中（图7）。

4.2 第二节：处处皆欢景

第二节的剧情进入慢节奏，"主角"（赏画者的思绪）正式进入画面，化身成为岸边的市民之一。她在两岸的各个地方游走，透过各个画中角色来体察龙舟即将穿桥而来的快乐瞬间。这一节是本片的重点（图8）。

《端阳》画面正中间的小岛、正上方的宅院、左下角的商铺群、右上角的水岸酒楼以及两处连桥是主要的"造景"内容。在短片的场景重构中，这一部分的人、桥、建筑等主景元素被保留，只将山、石、树等不必要的配景剥离，根据新的情节需要再重组进陆地和水面场景之中（图9）。

《端阳》画面空间大体分为三层：近景汀岸，中景水体，远景山林，结构清晰分明。水岸和桥对角呼应，水虚岸实，以桥相连。两座桥由右下到左中再到右上，水岸交替出现，从宁静到喧闹再归于宁静。这个节奏也是短片空间顺序的确定点和情节叙事的转折点。而"游走"的路径即由画面右下角的木桥作为入画的开篇，顺序依次为桥—左岸—桥—水—小岛—水—右岸。由此，第二节短片的场景依次为：木桥入

图 6 短片第一节部分内容截图

14

画（原图中右下赶场）—庭院驻足—商铺问询—上桥观景—登船入岛—上岸离去，共六组剧情。

"主角"走入画中，遇到一些人，看到一些风景，心情也起伏不定。这种心态的细微且悠然的变化与前一段的集体癫狂完全不同。她从桥上穿过蜿蜒的水面，感到迷茫和不安；沿着水岸来到一处人家，见此处树林掩映，在本该热闹的氛围中较显神秘，不禁感到疑惑；穿过树林路过街边商铺，向画中一老人询问当下境况，听闻秦淮河上龙舟赛事正酣，不由好奇；随即来到桥头，上桥见人头攒动，水面却只有一两只龙舟感到十分不解；于是乘船寻找，路过沙渚，登岛探寻后又乘船上岸；在一酒肆遇一人，交谈告别后离去。从入画到出画，结合路径设计、场景重构，在"主角"经历迷茫-不安-怀疑-好奇-不解-领悟六个阶段的情绪变化后，第二节剧情落幕。

在短片中，主角衣袖摇摆，款款而行，有几处情节做镜头放大停留处理，意指"主角"在与画中人有所交流，获得某些信息。虽然一路行来，左顾右盼，但从头到尾都没有看到大家期待中的"龙舟"。不过，即使"龙舟"胜景隐于画外，"主角"游走时，和着音乐节拍时而缓行、时而滑动、时而跳跃的步伐节奏踏过的一块块"小空间"却一致地显出快乐的气息，似乎每一栋房子、每一棵树、每一块土坡，都在压抑着兴奋之情，等待着某一刻的到来（图10）。

4.3 第三节：梦逝水长情

"主角"在画中体验完后，来到500年后的"现实"秦淮，两个时空体无声地叠合在一起（图11）。岁月匆匆，江南依旧，秦淮河水从古流到了今，但今之秦淮，已非过往，曾经的诗酒风流、满河沸腾的景象一去不返。从画中到画外，这是第三节即"终章"的剧情，也是作者构建给观众的思考空间（图12）。

剧本文字内容（部分）："她，依旧衣袖摇摆，款款而行。当她走在岸边步道上，她会发现，路很平坦干净，但是离水边似乎有点远。很多地方筑起了高高的堤坝，很难走到河边弯腰触碰一下河水。她遇到三三两两的人，大部分都是老人。她们悠闲地踱步。他们在各个空地处下棋、遛鸟、做运动，看着像是每天都常驻于此。他们似乎是今天秦淮河的'主人'。偶尔会有年轻的母亲带着小孩会闲逛。年轻人比较少，并且大多聚集在文德桥附近——那里小吃很多很热闹。贡院、夫子庙一如既往的喧哗，大家摩肩接踵，来去如梭。"

"她登上一艘游船。船比以前的窄多，虽然不是木头的，但也很整洁。船在河上缓缓前行，感觉河道窄了不少，沙渚也没有了。河岸很规整，没有过去那样'随意自然'。贡院对面的那堵墙还在，只是墙的颜色从白色变成红色，还多了两条巨龙。游船漂到一座桥下。桥的样子很陌生，但也能看出它似乎就是之前的文德桥（桥上刻有'三元及第'）。三拱还在，只是桥身粗壮了不少。'这样的秦淮河，似乎很难再赛龙舟了——龙舟在这个尺度的河道上会很挤，说不定还会撞上桥墩。'她

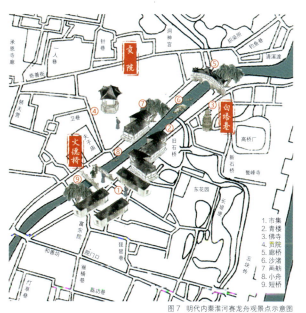

图7 明代内秦淮河赛龙舟观景点示意图

图8 短片第二节部分内容截图

图9 《端阳》中被保留的"景点"

15

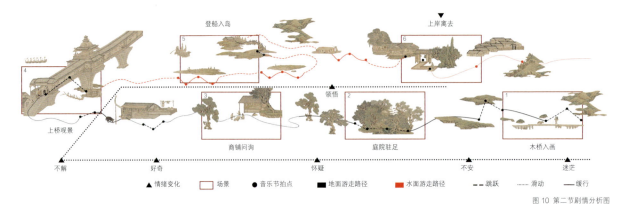

图 10 第二节剧情分析图

心里嘀咕了一下。还有，那个曾经的极度'奢华'的廊桥今何在？"

5. 结语

让古代图像"动起来"，是这次创作的主要路径。在古代，"影像"表达即是通过将图像简单动态化处理来实现的。最早的影像制作尝试可追溯到秦晋时期的走马灯——一种利用物理学冷热空气对流的原理带动叶轮旋转的花灯，叶轮转动时，可以呈现特定图案的动态效果。而真正意义上将图像结合光影成像技术、声音、情景以形成完整幕影效果的是始于汉代的皮影戏。本文的影像制作灵感就源于此种图像化的影像制作形式。

将古画"影像化"，即以对画面的图像分析为前提，再以此为基础，结合数字技术进行"动态化处理"。我们对《端阳》的图像分析从两个角度进行：一方面分析古画画面中的人物、情节等内容要素完成剧情设计；另一方面分析画面构图、造景等空间手法以重构影像场景。具体操作分为四步：（1）古画图像分析，包括历史背景梳理、画面内容解读、空间构图分解、创作心理推测等。（2）原址勘察，根据古画内容和其他佐证史料分析确定画面空间所对应的当下原址，了解其现存的空间状态。（3）移情重构，构思剧情，结合古画元素、布置具有古典空间美学的剧情场景。（4）将画面、原址二种空间元素融入影像中，实现时间和空间上的延伸，并结合配乐等完成影像制作。

再现古代图像中的"古意"，是这次创作的最终目的。在这个过程中，我们发现画中的秦淮河两岸、楼阁亭台极具美感，这些景象虽都已消失在现代城市中，但水陆基础结构尚在。我们认为，通过古画与城市中尚存的地理原址，再现那些空间美学是可能的。在剧本中，角色的路径和场景重构了剧情的三维空间，而"主角"的穿越串联起的古今对照则为影像叠加了第四个维度——时间。至此，一次从二维图像（古画）到四维影像（短片）的跨越式创作就完成了。

图 11 《端阳》平面与原址平面

图 12 短片第三节部分内容截图

1. The Qinhuai River and the Dragon Boat Festival

Since ancient times, the Qinhuai River has been one of the most representative spatial symbols of Nanjing. In the middle and late Ming Dynasty, the prosperity of the Qinhuai River reached a peak. Especially in the inner Qinhuai area of the city (mainly the south side of the river), countless literati and tourists lingered there. During the day, it is bustling and noisy, and at night there were sounds of oars, the shadows of lanterns and a faint fragrance in the air. The most lively day of the year on the Qinhuai River was the Dragon Boat Festival, when the people of the city gathered on the riverbank to hold the most festive group activity—the dragon boat race. Ming's *Zhengde Jiangning County Chronicles* contains: "People buy boats and carry wine to play, and the custom is the most prosperous in recent years." In addition, at the end of the Ming Dynasty and the beginning of the Qing Dynasty, Zhang Dai's *Dream Memories of Tao'an* recorded the grand occasion of the race: "I watched the dragon boat races on the West Lake twelve or thirteen times, Jisi race on the Qinhuai River, Xinwei race in Wuxi, Renwu race at Guazhou, at Jinshan Temple."

To this day, it is common to see written records of dragon boat racing on the Qinhuai River in the Ming Dynasty. What's even more rare is that there is also a painting titled *Duanyang (Dragon Boat Festival)* depicting this scene that has been handed down to the world, which can make future generations feel the most prosperous urban space (Qinhuai River) more intuitively, which coincides with the climax of the festival atmosphere (Dragon Boat Festival).

2. Ancient painting *Duanyang (Dragon Boat Festival)*

Duanyang (Dragon Boat Festival) is a set of colored paper in *Suihua Jisheng Tu (Pictures Recording the Prosperity of the Seasons)*, one of the documentary albums painted by the Ming Dynasty painter Wu Bin (Figure 1). This group of pictures is based on the theme of the twelve solar terms, showing the prosperous scene of a peaceful and prosperous era. Although there is no specific inscription on the painting, there is almost a consensus among researchers that the scene in the painting is supposed to be taken from the city of Nanjing during the Ming Dynasty, around 1595–1612. In addition to the realism of the pictures and the delicate techniques, this group of pictures has another ingenious concept. The 12 paintings are set up according to the special festivals of each month respectively, and if they are arranged in chronological order, you can get a glimpse of the time flow and the changes of the four seasons in the Ming Dynasty. If each picture is further associated with a movie plot, the social activities of different seasons and the urban landscapes of different locations are vividly presented in one "scene" after another.

The length-to-width ratio of the painting is about 1 : 2, the viewing angle is axonometric, the water body is the center, and the two sides of the river are obliquely paired from the lower left to the upper right, which are connected by bridges. The water-covered bridge, the sandbank in the water, and the sails of the boat on the water form the main part of the picture, and the eyes of the characters on the bridge converge on the water, and one or two dragon boats can be seen hidden between the swaying leaves. In the Ming Dynasty, dragon boat races were generally held in both the inner and outer Qinhuai River. The outer Qinhuai River section is in front of the Porcelain Tower and Wuyi Lane, and the river continues to the vicinity of Changgan Bridge and Zhonghua Gate. The inner Qinhuai River section is crossed from Qingxi Ferry to the Imperial Examination Hall and the area around the Confucius Temple (Figure 2). The main "attraction" in *Duanyang (Dragon Boat Festival)* is the covered bridge that crosses the river. The design of this covered bridge is extremely gorgeous. The bridge body is long, and the piers are curved inward, which is very elegant. In the middle of the bridge is a two-storey "tower" with a "Buddha shrine" inside, where the tall Buddha statue and the surrounding circle of small Buddha statues are clearly visible (Figure 3). It is unlikely that such an elaborate and luxurious covered bridge could have appeared on the Outer Qinhuai River, because the bridges of the Outer Qinhuai River basically have the function of defending against war, and their forms are very simple. Therefore, according to the preliminary analysis, the scene in the painting should be somewhere on the Qinhuai River at the existing Confucius Temple and the east end of Wende Bridge. The ornate covered bridge should have been the "Lishe Bridge" in front of the Huiguang Temple on the south bank of the river—now it is the Wenzheng Bridge. *Duanyang (Dragon Boat Festival)* is supposed to a picture depicting the dragon boat race on the banks of the Inner Qinhuai River in the late Ming Dynasty.

The "Qinhuai and Dragon Boat Festival" drama staged in *Duanyang (Dragon Boat Festival)* kicked off—In the middle of the picture, the small island and the scene of people boarding the beach by boat to worship the temple, watch the races, and enjoy the scenery, form a scene full of wild and interesting; in the lower left corner of the picture, the compact shops next to the bridgehead are busy shops and customers buying goods to form a lively market buying and selling scene; in the upper left covered bridge, the spectators of the dragon boat race are cheering while leaning on the railing, and curious pedestrians stop to look; the upper right corner is densely forested, with tea houses along the coast and village houses in the distance, creating a quiet and wild atmosphere; in the lower right corner of the picture, a man with a monkey on his back seems to be eager to catch up with the excitement, taking the opportunity to perform juggling and make a lot of money. "After lunch, people go to watch the dragon boat race one after another." The scene of shop trading, watching the race at the bridge, the island outing, the mountain forest wine house and other quiet miscellaneous places are presented in front of you in turn.

3. From painting to image: A visual experiment in *Dragon Boat Festival 2023*

The short film *Dragon Boat Festival 2023* is an attempt to visualize ancient paintings based on historical research, combined with image analysis and digital technology. The focus of the finished image is to respond to the "meaning" in the ancient painting.

According to the above analysis, the location of the original site in the picture space of *Duanyang (Dragon Boat Festival)* is easy to determine, but the "ancient meaning" in the painting is somewhat obscure. The bird's-eye view of the painting with the water as the center and the bridge as the commanding height of the space is obviously to cope with the hot scene of the dragon boat race on the water, but the dragon boat as the "protagonist" is extremely inconspicuous in the painting—a dragon boat is hidden between the branches of a willow tree on the left side of the picture, and another is hidden behind a huge bridge pier, revealing only two rowers, three red oars and a small section of the hull. If you don't pay attention and search carefully, they will hardly be found. The spectators on the bridge and the crowd on the shore of sand are crowded, they either probe or twist their necks, but the place where they gaze is empty, and the atmosphere is slightly confusing (Figure 4). This is undoubtedly the unique setting of the painter Wu Bin, the fiery scene has not yet appeared, and everyone is waiting for the moment when the first dragon boat rushes out under the bridge. This may be the fun of the "dragon boat race" that only the truly immersive ancients understood.

Another hidden "ancient meaning" is the living conditions on both sides of the river. The lower side of the river is lined with shops and taverns, while the houses on the upper side of the river are much more elaborate. At the back of some of the houses are sizable courtyards. They stretched out from the shore and sat in front of the windows, all of them yet women, they were beautifully dressed, and were chatting and watching the scenery with their fans (Figure 5). According to the situation at that time, the courtyards with overlooks by the river belonged to the category of "old courtyards" (brothels) . Those women looking out of the window are also part of the legend of Qinhuai courtesans. In the current reality, these "ancient meanings" have not left a trace. Except for the fact that the Qinhuai River is still flowing, everything has disappeared. It seems that it is not easy to recreate that "ancient meaning".

4. Script: "Exploration"

The script intends to follow the rhythm of analysis of first seeing its appearance, then observing its shape, and finally revealing its meaning—with the viewer's thoughts as the "protagonist" and "exploration" as the clue, the plot unfolds along the structure from ancient times to the present, from inside the painting to outside the painting. The whole play has three acts: at the beginning, the lively scene of the Dragon Boat Festival is presented, which corresponds to the intuitive feeling when first seeing the picture, and focuses on rendering the joyful atmosphere climax; then, the thoughts went into the painting, wanting to "explore" the fun in the painting in person; after coming out of the painting, returning to the "reality" River—the current Qinhuai River, and to "search" for the traces of memory 500 years ago.

4.1 Act 1: Dragon boats on the Qinhuai River

The "protagonist" at the beginning was infected by the lively scene, and unconsciously incarnated as a dragon boat rower, shaking the oars with everyone, sweating and passionate. The scenes are simulated using digital modeling technology, and the perspective is constantly switched between the vicinity of the galloping dragon boats and the applauding audience on the shore. This part of the lens requires multiple-angle changes, a sense of speed, close-ups, and cross-cutting of large space overlooks (Figure 6).

Another important point in this scene is the peculiarity of the background space. The spatial form of this section of the Inner Qinhuai River is quite rich, and it has a strong Nanjing character. In addition to the covered bridge we see in *Duanyang* (*Dragon Boat Festival*)l, there are also the terraces of the river house, the shore, the sand island, the tributary wooden bridge, and the riverside boat. The large flat "square" in front of the Imperial Examination Hall can also accommodate a large number of people. The Huiguang Temple by the river may also have high-viewpoint viewing places such as a two-story pavilion and a tower.

In detail, there are about 10 different places to watch the dragon boat race. These staggered viewpoints (bazaars, brothels, Buddhist temples, tribute courtyards, covered bridges, sand island, painting boats, small boats, and short bridges) are all merged into the wild "opening" of the "Dragon Boat Drama in Qinhuai" (Figure 7).

4.2 Act 2: Happy scene everywhere

In the second act, the plot moves into a slow pace, and the "protagonist" (the viewer's thoughts) officially enters the picture and becomes one of the citizens on the shore. She travels around the two sides of the river, and through the characters in the painting, she perceives the happy moment when the dragon boat is about to pass through the bridge. This act is the focus of the film (Figure 8).

The island in the middle of *Duanyang* (*Dragon Boat Festival*), the house directly below, the shops in the lower-left corner, the waterfront restaurant in the upper-right corner, and the two bridges are the main "landscaping" contents. In the scene reconstruction of the short film, the main scenic elements such as people, bridges, and buildings in this part are retained, and only the unnecessary background scenery such as mountains, rocks, and trees are stripped away, and reorganized into land and water scenes according to the needs of the new plot (Figure 9).

The picture space of *Duanyang* (*Dragon Boat Festival*) is roughly divided into three layers: a close-up view of shore, a mid-view water body, and a distant view of mountains and forests, with a clear and distinct structure. The waterfront and the bridge echo diagonally, and the water bank is actually connected by a bridge. The two bridges alternate from the lower right to the middle left to the upper right, and the waterfront appears alternately from quiet to noisy and back to quiet. This rhythm is also the determining point of the spatial order of the short film and the turning point of the plot narrative. The path of "wandering" begins with the wooden bridge at the lower-right corner of the painting, and the order is bridge – left bank – bridge– water – small island – water – right bank. As a result, the second act of the short film is as follows: entering the painting by the wooden bridge (the lower right in the original picture) – stoping at the courtyard – inquiring at the shop – going up the bridge to view the scenery – boarding the boat and entering the island – going ashore and leaving, a total of six sets of plots.

The "protagonist" walks into the painting, meets some people, sees some scenery, and her mood fluctuates. This subtle and leisurely change in mentality is completely different from the collective madness of the previous determining. She crossed the winding water from the bridge, feeling lost and uneasy; along the waterfront, she comes to a house, and sees that it is hidden by the woods here and is more mysterious in the atmosphere that should be lively, and can't help but wonder; walking through the woods and passing by the street shops, she asks an old man in the painting about the current situation, and she can't help but be curious when she hears that the dragon boat race on the Qinhuai River is in full swing; then she comes to the bridge head, and when she goes up the bridge, she sees the crowd, but there are only one or two dragon boats on the water, and she is very puzzled; so she takes a boat to look for it, passes by the sand island and lands on the island to explore, and then goes ashore by boat; she meets someone at a tavern, has a conversation, says goodbye and leaves. From entering the painting to exiting the painting, combined with path design and scene reconstruction, the second act ends after the "protagonist" experiences six stages of emotional changes of confusion, anxiety, doubt, curiosity, incomprehension, and comprehension.

In the short film, the protagonist's sleeves are swaying, and there are several plots where the camera is zoomed in and stopped, which means that the "protagonist" is communicating with the people in the painting and obtaining certain information. Although she looks around along the way, she didn't see the "dragon boat" that everyone was expecting from beginning to end. However, even though the scenery of the "dragon boat" is hidden outside the painting, the "small spaces" that the "protagonist" walks through to the rhythm of the music, sometimes slowing, sometimes sliding, and sometimes jumping, show a happy atmosphere, as if every house, every tree, and every dirt slope is suppressing excitement and waiting for a certain moment to arrive (Figure 10).

4.3 Act 3: Dreams fade while the river flows with enduring sentiments

After the "protagonist" is experienced in the painting, she comes to the "reality" of Qinhuai 500 years later, and the two time-space bodies are silently superimposed together (Figure 11). Time has passed in a hurry, the south region of the Yangtze River is still the same, and the water of the Qinhuai River has flowed from ancient times to the present, but today's Qinhuai is no longer what it used to be, and the once romantic scenes of poetry, wine and the bustling river are gone. From the inside of the painting to the outside of the painting, this is the plot of the third act, which is also the "final chapter", and it is also the space that the author constructs for the audience to think about (Figure 12).

The text of the script (partial): "She is still walking gracefully with her sleeves swaying. When she walks on the shore path, she will find that the road is flat and clean, but it seems to be a little far from the water's edge. In many places, high dikes have been built, making it difficult to walk to the river bank and bend down to touch the water. Most of the people she met were old people pacing leisurely. They play chess, walk birds, and play sports in various open spaces, and it looks like they are there every day. They seem to be the 'masters' of the Qinhuai River today. Occasionally, young mothers come in with their little ones to hang out. There are fewer young people, and most of them gather around Wende Bridge—where there are plenty of snacks and a lot of fun. The Imperial Examination Hall and Confucius Temple were as noisy as ever, and people were rubbing shoulders and coming and going."

"She boards a pleasure boat. The boat is wider than before, and although it was not wooden, it is also neat. The boat slowly moves forward on the river, feeling that the river has become much narrower, and the sand islands are gone. The riverbank is very regular, and it is not as 'random and natural' as it used to be. The wall opposite the Imperial Examination Hall is still there, but the color of the wall has changed from white to red, and there were two more dragons. The boat floats under a bridge. The appearance of the bridge is very unfamiliar, but it can also be seen that it seems to be the former Wende Bridge (the bridge is engraved with 'Sanyuan Jidi' (won the first place in the township examination, the regional examination, and the palace examination). The three arches are still there, but the bridge is much sturdier. 'It seems that it is difficult to race dragon boats in the Qinhuai River like this—the dragon boats will be very crowded on the river at this scale, and they may even hit the bridge piers.' She muttered to herself. Also, where is the once extremely 'luxurious' covered bridge?"

5. Conclusion

Making ancient images "move" is the main path of this creation. In ancient times, the expression of "image" was achieved by simply moving the image. The earliest attempts at image-making can be traced back to the Qin and Jin dynasties, a kind of lantern that used the principle of convection of hot and cold air in physics to drive the rotation of the impeller, and when the impeller rotated, the dynamic effect of a specific pattern could be seen. In the true sense, shadow puppetry began in the Han Dynasty by combining images with light and shadow imaging techniques, sounds, and scenes to form a complete screen effect. The inspiration for this article is derived from this form of pictorial video production.

The "visualization" of ancient paintings is based on the premise of image analysis of the picture, and then combined with digital technology to carry out "dynamic processing". We analyze the image of *Duanyang* (*Dragon Boat Festival*) from two perspectives: on the one hand, we analyze the characters, plots and other content elements in the ancient paintings to complete the plot design; on the other hand, we analyze spatial techniques such as picture composition and landscaping to reconstruct the image scene. The specific operation is divided into four steps: (1) image analysis of ancient paintings, including historical background combing, interpretation of picture content, spatial composition decomposition, creative psychological speculation, etc.; (2) in-situ site investigation, according to the content of ancient paintings and other supporting historical materials, determining the current original site corresponding to the picture space, and understanding its existing spatial state-based; (3) empathy-based reconstruction, conception of the plot, combination of ancient painting elements, and arrangement of plot scenes with classical space aesthetics; (4) the two spatial elements of the surface and the original site are integrated into the image to realize the extension of time and space, and the image production is completed in combination with the soundtrack.

Reproducing the "ancient meaning" in ancient images is the ultimate goal of this creation. In this process, we found that the two banks of Qinhuai and the pavilions in the painting are very beautiful. Although these scenes have disappeared in the modern urban development, but the water and land infrastructure is still there. We believe that it is possible to reproduce those spatial aesthetics through ancient paintings and the remaining geographical sites in the city. In the script, the paths and scenes of the characters reconstruct the three-dimensional space of the plot, and the contrast between the ancient and the modern connected by the passage of the "protagonist" superimposes a fourth dimension—time on the image. At this point, a leapfrog creation from two-dimensional images (ancient paintings) to four-dimensional images (short films) has been completed.

课程概览
COURSE OVERVIEW

建筑通史 GENERAL HISTORY OF ARCHITECTURE

建 筑 通 史
GENERAL HISTORY OF ARCHITECTURE

王 骏 阳
WANG Junyang

课程简介

作为一门建筑与城规实验班本科一年级学科基础认知课程和外专业通识选修课程，建筑通史并不仅是传统意义上从古代到当代的逐个历史分期的"通"，也是打破中外建筑史和中西建筑史学科隔阂的"通"，是古今中外的"通"，是建筑史与建筑学概论的"通"，是建筑学专业认知与人类文明和文化历史的"通"。本课程的教学旨在培养学生立足本土、胸怀世界的人文情怀和广博的专业知识技能。

课程内容

第一讲　走出建筑与建筑物的悖论
第二讲　居住与建筑的起源与发展（一）
第三讲　居住与建筑的起源与发展（二）
第四讲　神秘的人类早期文明
第五讲　融合与冲突中的西亚与伊斯兰建筑
第六讲　古希腊和古罗马建筑
第七讲　欧洲中世纪建筑
第八讲　意大利文艺复兴建筑及后续
第九讲　现代建筑
第十讲　密斯·凡·德·罗、勒·柯布西耶、阿道夫·路斯：现代建筑的三种道路
第十一讲　后现代建筑
第十二讲　日本建筑
第十三讲　世界建筑史中的中国建筑与西村大院
第十四讲　结构、空间、形式

Course introduction

As a first-year undergraduate suject-based course of Architecture and Urban Planning Experimental class and a general elective course for non-major students, General History of Architecture is not only a traditional "general" course from ancient times to contemporary times, but also a "general" course that breaks down the disciplinary boundaries between Chinese and foreign architectural history and Chinese and Western architectural history, a "general" course between ancient and modern times, a "general" course between the history of architecture and introduction to architecture, and a "generalization" of the professional knowledge of architecture and the history of human civilization and culture. The aim of this course is to cultivate students humanistic sentiment of having a global vision while being rooted in the local community, as well as broad professional knowledge and skills.

Course content

Lecture 1　Out of the Paradox of Architecture and Buildings
Lecture 2　The Origin and Development of Habitat and Architecture (I)
Lecture 3　The Origin and Development of Habitat and Architecture (II)
Lecture 4　The Mysterious Early Human Civilization
Lecture 5　West Asian and Islamic Architecture in the Context of Integration and Conflict
Lecture 6　Ancient Greek and Roman Architecture
Lecture 7　European Medieval Architecture
Lecture 8　Italian Renaissance Architecture and Its Aftermath
Lecture 9　Modern Architecture
Lecture 10　Ludwig Mies van der Rohe, Le Corbusier, Adolf Loos: Three Paths of Modern Architecture
Lecture 11　Postmodern Architecture
Lecture 12　Japanese Architecture
Lecture 13　Chinese Architecture and the West Village Basis Yard in the History of World Architecture
Lecture 14　Structure, Space, Form

课程作业答题选登

名词解释：造宫之法

"造宫之法"最早出现于1864年出版的《华英英华辞书》中，是对单词"architecture"的中文翻译。其中"宫"不是指后来的"宫殿"，而是房屋的统称，对应了"宫谓之室，室谓之宫"，"宫"是房屋和院落的统称，"室"是房屋单元的称呼。"造宫之法"反映了根据中国传统文化对西方"建筑"一词进行的理解。

黎宇航，2021 建筑学

名词解释：新建筑五点

"新建筑五点"是法国建筑师勒·柯布西耶在1926年提出的观点，主要包括：底层架空支柱、屋顶花园、自由平面、横向长窗、自由立面。底层架空支柱指用支柱使得建筑底层远离地面架空；屋顶花园指将花园设置在房屋顶部；自由平面指对平面的选择可以按需求决定；横向长窗指在建筑侧面水平安置长条窗，以扩大视野增加采光；自由立面指从立面上看建筑不受空间限定，充满自由性。

"新建筑五点"是基于多米诺体系提出的，它将板柱体系作为承重结构。柯布西耶1920年的蔚蓝海岸住宅方案设计是"新建筑五点"的雏形。他于1926年正式提出"新建筑五点"，为现代主义建筑的发展奠定了基础。

刘超然，2022 工科实验班

名词解释：阿卡·汗建筑奖

阿卡·汗建筑奖始于1970年代，用于每三年颁发评选一次对伊斯兰建筑有贡献的建筑师、建筑项目及具体单体作品。该奖项并不偏执于对伊斯兰建筑原始或经典传统的继承，而是着重关注建筑师对伊斯兰建筑风格的发展，并受现代建筑风格潮流的影响，将建筑本身对人与自然环境的调和优劣作为评选的重要参考。例如，印度尼西亚机场继承了伊斯兰建筑较原始传统的连排草顶木屋的外观设计，而半开放式的格局，让机场室内环境宜人，契合印度尼西亚地区热带雨林的气候，实现了人与环境及当地特色文化的协调，因此获得了阿卡·汗建筑奖。

王玉欣，2022 建筑与城市规划实验班

名词解释：倒置悬链

"倒置悬链"来源于一种物理现象，即固定一条铁链的两端，让中部铁链自由下坠形成弧形。罗伯特·胡克率先提出将这种现象运用到教堂穹顶设计中。他认为，假设铁链中间的重量为穹顶自重，再将产生的弧形倒转，便是穹顶可拥有最自然合理的受力状态。然而，"倒置悬链"得到的穹顶形状太尖，这导致圣保罗大教堂在"倒置悬链"穹顶外搭建形状更加饱和的穹顶，内部则以另一个内壳形成空间。从这个案例可以看出，好的结构形式可能不够美观，而美观的形式也许不是最合理的结构。因此需要建筑师权衡二者的关系，实现二者的矛盾统一。

张耀天、王明宇，2022 建筑与规划实验班

Selected coursework answers

Terminology: The Method of Building a Gong

The term "The Method of Building a Gong" first appeared in the 1864 edition of the *Chinese-English and English-Chinese Dictionary*, and is the Chinese translation of the word "architecture". The word "Gong" does not refer to the later term "palace", but is a generic term for houses, corresponding to the phrase "the palace is called a room, the room is called a palace", and "Gong" is the name for houses and courtyards, and "room" is the name of the housing unit. "The method of building a Gong" reflects an understanding of the Western term "architecture" based on traditional Chinese culture.

LI Yuhang, 2021 Architecture

Terminology: Five Points of New Architecture

The "Five Points of the New Architecture", proposed by French architect Le Corbusier in 1926, mainly include: pilotis, roof gardens, free plan, long horizontal windows, and free facade. Pilotis refers to using pillars to elevate the ground floor of the building away from the ground; roof gardens refer to the garden that will be set on the top of the house; free plan refers to the choice of planing that can be decided according to the demand; horizontal long windows refer to installing the long window horizontally on the side of the building, expanding the field of view and increasing the light; free elevation refers to the elevation of the building that is not limited by space, full of freedom.

The "Five Points of New Architecture" are based on the domino system, with the slab-column system as the load-bearing structure. The design of the Villa on the Côte d'Azur by Le Corbusier in 1920 was the prototype of the "Five Points of New Architecture". He formally put forward the "Five Points of New Architecture" in 1926, which laid the foundation for the development of modernist architecture.

LIU Chaoran, 2022 Engineering Experimental Class

Terminology: Aga Khan Award for Architecture

The Aga Khan Award for Architecture was initiated in the 1970s and is awarded every three years to recognize architects, architetural projects and individual works that have contributed to Islamic architecture. The award does not emphasize on inheriting the original or classical traditions of Islamic architecture, but focuses on the development of Islamic architectural styles by architects. It is influenced by modern architectural trends, and takes the quality of how the building itself reconciles human beings and the natural environment as an important reference. For example, the airport in Indonesia inherited the more primitive and traditional Islamic architectural design of rows of grass-roofed wooden houses, while the semi-open layout of the airport allows for a pleasant indoor environment that suits the climate of the Indonesian region's tropical rainforests, realizing the harmony between human beings and the environment as well as the local characteristic culture. Thus, it won the Aga Khan Award for Architecture.

WANG Yuxin, 2022 Architecture and Urban Planning Experimental Class

Terminology: Inverted Suspension Chain

The "Inverted Suspension Chain" is derived from a physical phenomenon in which the ends of a chain are fixed so that the center chain is free to drop down to form an arc. Robert Hooke pioneered the use of this phenomenon in the design of church domes. He argued that assuming the weight in the middle of the chain was the self-weight of the dome, and then inverting the resulting arc, the dome would have the most natural and reasonable state of stress. However, the shape of the dome obtained by the "Inverted Suspension Chain" was too pointed, which led St. Paul's Cathedral to build a more saturated dome outside the "Inverted Suspension Chain" dome, with another inner shell forming the space inside. From this case, it can be seen that a good structural form may not be aesthetically pleasing enough, and an aesthetically pleasing form may not be the most rational structure. Therefore it is necessary for the architect to balance the relationship between the two and achieve the contradictory unity of the two.

ZHANG Yaotian, WANG Mingyu, 2022 Architecture and Planning Experimental Class

建成环境导论与学科前沿
INTRODUCTION TO THE BUILT ENVIRONMENT AND FRONTIERS OF THE DISCIPLINE
丁沃沃
DING Wowo

课程简介

建成环境是人类生产与生活的基本场所，是生存和发展的重要环境。建成环境的优劣关系到每个人的生存状况，建成环境的构建涉及多个学科，与它相关的各类知识是多个学科的共同基础。有史以来，一方面，人们依靠技术进步不断从地球获取资源，同时也不断创造出更加高效和舒适的生存空间；另一方面，随着对自然界认知的更新，人们也在不断调整建构建成环境的方法和路径。

因此，从专业的角度了解建成环境的概况、进展、问题和前景对于刚刚踏入本学科的初学者来说是后续学习的基础知识。此外，本课程力图承担训练大学学习方法的任务，以教学过程为载体，引导学生借助新的媒体技术获取知识，培养独立思考、思辨的能力，促进学生尽快完成从高中到大学学习方法的转型，为今后的学习打好基础。

课程要求

1）理解随着社会转型，城市建筑的基本概念在建筑学核心理论中的地位以及认知的视角。

2）通过理论的研读和案例分析理解建筑形式语言的成因和逻辑，并厘清中西方不同的发展脉络。

3）通过研究案例的解析理解建筑形式语言的操作，并掌握设计研究的方法。

Course introduction

The built environment is the basic place for human production and life, and an important environment for survival and development. The quality of built environment is related to everyone's living conditions. The construction of built environment involves many disciplines, and various kinds of knowledge related to it are the common foundation of many disciplines. Historically, on the one hand, people rely on technological progress to constantly obtain resources from the earth, and also constantly create more efficient and comfortable living space; on the other hand, with the renewal of people's understanding of nature, they are constantly adjusting the methods and paths of constructing built environment.

Therefore, understanding the general situation, progress, problems and prospects of the built environment from a professional perspective is the basic knowledge for beginners who have just entered the discipline. In addition, this course aims to undertake the task of training university learning methods, using the teaching process as a carrier to guide students on how to use new media technologies to acquire knowledge, cultivate independent thinking and critical thinking abilities, and promote students to complete the transition from high school learning methods to university learning methods as soon as possible, laying a solid foundation for their future learning.

Course requirements

1) To understand the status and cognitive perspective of basic concept of urban buildings in the core theory of architecture with the social transformation.

2) To understand the reason and logic of architectural formal language and clarify different development processes in China and the West through theory reading and case analysis.

3) To understand the operation of architectural formal language and grasp methods of design study by analyzing research cases.

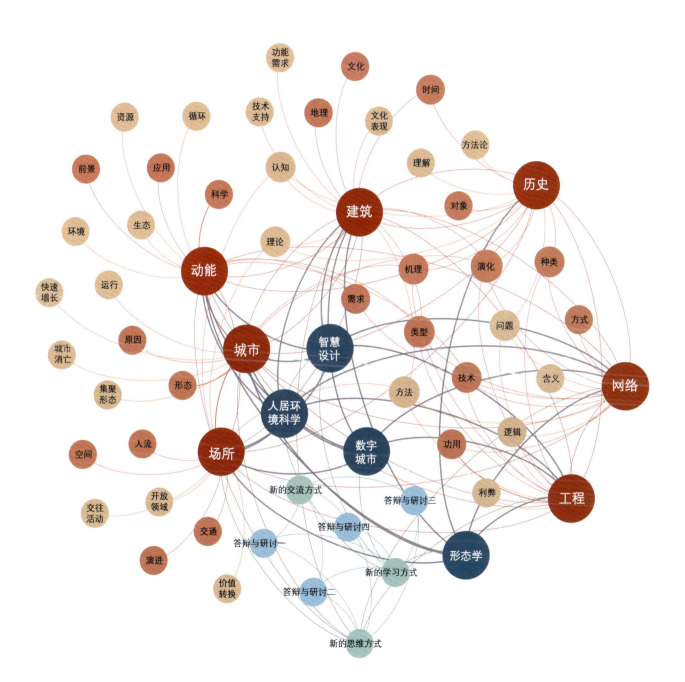

设 计 基 础
DESIGN FOUNDATION

刘铨 黄春晓 史文娟 梁宇舒
LIU Quan HUANG Chunxiao SHI Wenjuan LIANG Yushu

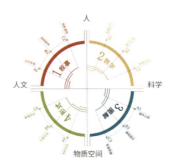

第一阶段：知觉、再现与设计
　　知识点：人与物——材料的知觉特征（不同状态下的色彩、纹理、平整度、透光性等）与物理化学特征（成分、质量、力学性能等）；摄影与图片编辑——构图与主题、光影与色彩；三视图与立面图绘制；排版及其工具——标题、字体、内容主次、参考线。

第二阶段：需求与设计
　　知识点：人与空间——空间与尺度的概念，行为、动作与一个基本空间单元或空间构件尺寸的关系；平面图、剖面图、轴测图的绘制；线型、线宽、图幅、图纸比例、比例尺、指北针、剖断符号、图名等的规范绘制。

第三阶段：制作与设计
　　知识点：物与空间——建构的概念；空间的支撑、包裹与施工；实体模型制作——简化的建造；计算机建模工具——虚拟建造；透视图绘制。

第四阶段：环境与设计
　　知识点：人、物与空间——城市形态要素、城市肌理与城市外部空间的概念；街道系统与交通流线；土地划分与功能分类；总平面图、环境分析图（图底关系、交通流线、功能分区、绿地景观系统）；照片融入表达。

Phase one: Perception, representation and design
　　Knowledge points: People and objects—the perceptual characteristics of materials (color, texture, flatness, light transmittance, etc. in different states) and physical and chemical characteristics (composition, quality, mechanical properties, etc.); photography and picture editing—composition and theme, light, shadow and color; three-view and elevation drawing system; layout and its tools—headings, fonts, primary and secondary content priority, and reference lines.

Phase two: Requirements and design
　　Learning points: People and space—the concept of space and scale, the relationship between behavior, action and the size of a basic spatial unit or spatial component; plan, section, and axonometric drawings; standardized drawings of line type, line width, map size, drawing scale, scale bar, compass, section symbol, drawing title, etc.

Phase three: Production and design
　　Learning points: Objects and space—the concept of construction; space support, wrapping and construction; physical model making—simplified construction; computer modeling tools—virtual construction; perspective drawing.

Phase four: Environment and design
　　Learning points: People, objects and space—the concept of urban form elements, urban texture and urban external space; street system and traffic flow; land division and functional classification; general plan, environmental analysis map (relationship between map and ground, traffic flow, functional zoning, green space landscape system); integration of images in expression.

人与物	人与空间	物与空间	人、物与空间
3周（个人作业）	3周（个人作业）	4周（个人作业）	4周（个人作业）
每组10人左右	每组10人左右	每组10人左右	每组10人左右

 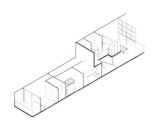  

A1 梁宇舒　　　　　　B1 梁宇舒　　　　　　　　C1 梁宇舒　　　　　　D1 梁宇舒
寻找木材　　　　　　马赛公寓抄绘及公寓设计　　逻辑生长　　　　　　聚落重构

A2 刘铨　　　　　　　B2 刘铨　　　　　　　　　C2 刘铨　　　　　　　D2 黄春晓
砖与砌块　　　　　　楼梯间　　　　　　　　　观景亭设计　　　　　观景亭选址与总平面设计

  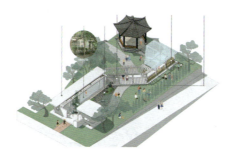

A3 史文娟　　　　　　B3 史文娟　　　　　　　　C3+D3 史文娟
双亭认知　　　　　　双亭改造　　　　　　　　榴园改造

27

设计基础 DESIGN FOUNDATION
模块A:寻找木材
MODULE A: LOOKING FOR WOOD

梁宇舒
LIANG Yushu

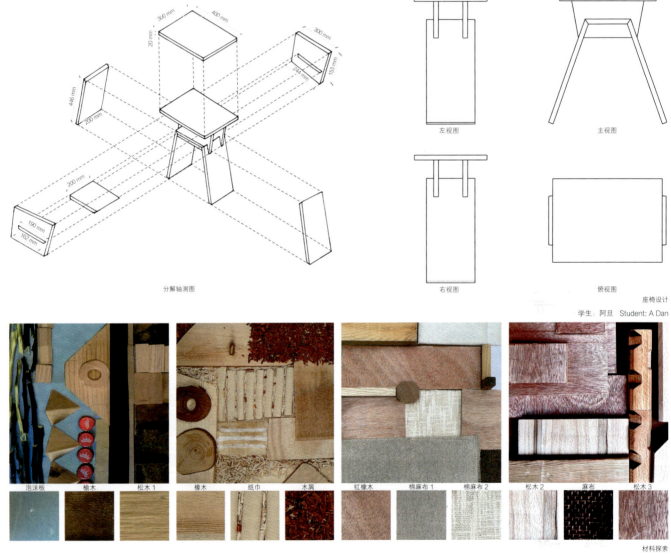

座椅设计
学生：阿旦　Student: A Dan

材料探索
学生：阿旦，张珺仪，扎西央宗，次旦央宗　Students: A Dan, ZHANG Junyi, TASHI Yangzom, TAETAN Yangzom

设计基础 DESIGN FOUDNATION

模块B:双亭改造
MODULE B: RENOVATION OF TWO PAVILIONS

史文娟
SHI Wenjuan

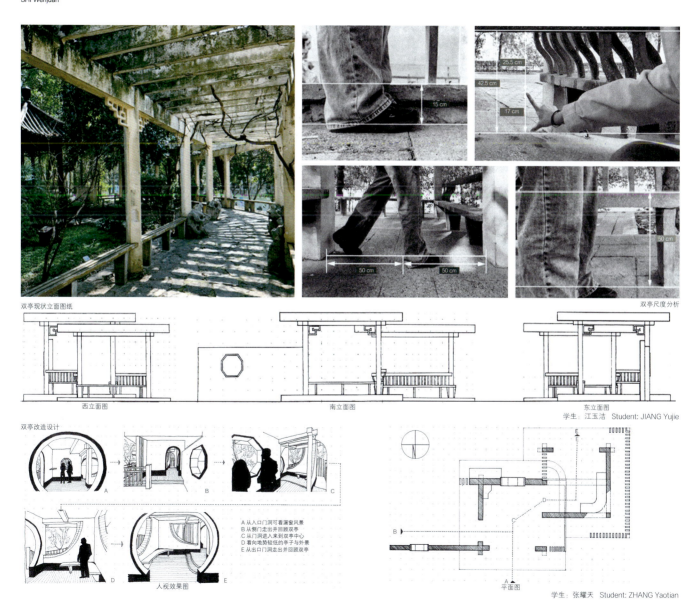

双亭现状立面图纸

双亭尺度分析

西立面图　　　　　　南立面图　　　　　　东立面图

学生：江玉洁　Student: JIANG Yujie

双亭改造设计

A 从入口门洞可看漏窗风景
B 从侧门走出并回顾双亭
C 门洞进入来到双亭中心
D 看向地势较低的亭子与外景
E 从出口门洞走出并回顾双亭

人视效果图

平面图

学生：张耀天　Student: ZHANG Yaotian

设计基础 DESIGN FOUDNATION

模块C:观景亭设计
MODULE C:OBSERVATION PAVILION DESIGN

刘铨
LIU Quan

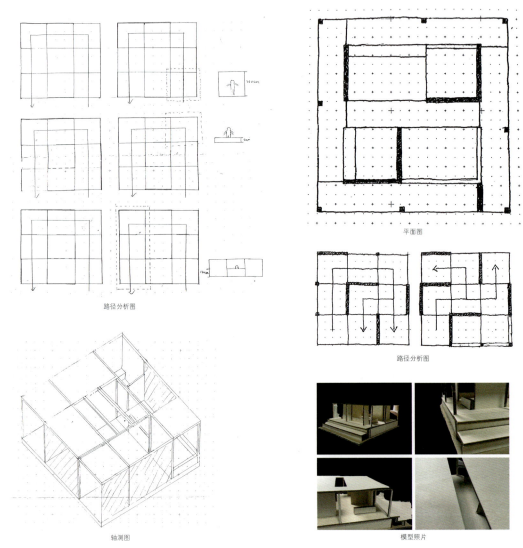

路径分析图

平面图

路径分析图

轴测图

模型照片

学生：张效儒　Student: ZHANG Xiaoru　　　学生：王馨柠　Student: WANG Xinning

模块D：榴园改造
MODULE D: POMEGRANATE GARDEN RENOVATION

史文娟
SHI Wenjuan

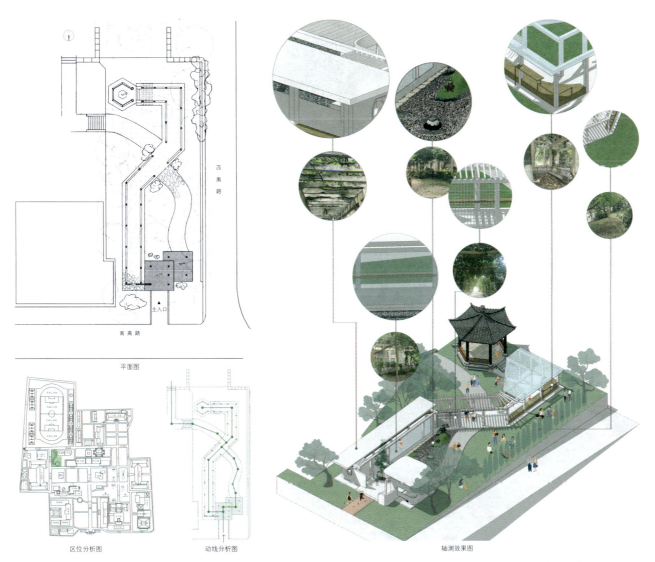

平面图

区位分析图　　动线分析图　　轴测效果图

学生：江玉洁　Student: JIANG Yujie

限定与尺度——独立居住空间设计
LIMITATION AND SCALE—INDEPENDENT LIVING SPACE DESIGN

刘铨　杨舢　唐莲　何仲禹　吴佳维
LIU Quan　YANG Shan　TANG Lian　HE Zhongyu　WU Jiawei

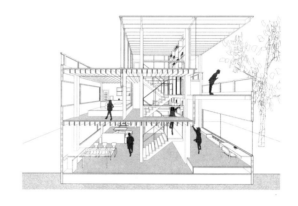

课程内容

本次练习的主要任务是综合运用前期案例学习中的知识点——建筑在水平方向上如何利用高度、开洞等操作划分空间，内部空间的功能流线组织及视线关系，墙身、节点、包裹体系、框架结构的构造方式，周围环境条件对空间、功能、包裹体系的影响等，初步体验一个小型独立居住空间的设计过程。

教学要点

1）场地与界面：本次设计的场地面积为 80—100 m^2，场地单面或相邻两面临街，周边为 1—2 层的传统民居。

2）功能与空间：本次设计的建筑功能为小型家庭独立式住宅（附设有书房功能）。家庭主要成员包括一对年轻夫妇和 1—2 位儿童（7 岁左右）。新建筑面积 160—200 m^2，建筑高度 ≤ 9 m（不设地下空间）。设计者根据设定的家庭成员的职业及兴趣爱好确定空间的功能（职业可以是但不局限于理、工、医、法的技术人员）。

3）流线组织与出入口设置：考虑建筑内部流线合理性以及建筑出入口与场地周边环境条件的合理衔接。

4）尺度与感知：建筑中的各功能空间的尺寸需要以人体尺度及人的行为方式作为基本参照，并通过图示表达空间构成要素与人的空间体验之间的关系。

教学进度

本次设计课程共 6 周。

第 1 周：构思并撰写几个有代表性的生活场景（家庭人物构成、人物相对关系、类似戏剧中的"折"之剧本）。四个地块分别制作 1：100（60 cm×60 cm）的场地模型。四个图纸及 1：100 手绘模型构思所用空间及其关系。

第 2 周：用 1：50 手绘平、立、剖面图纸，结合 1：50 工作模型辅助设计，在初步方案的基础上考虑功能与空间、流线与尺度。

第 3 周：确定设计方案，制作体现功能关系的空间关系模型（例如立体泡泡图）、推进剖、立面设计。

第 4 周：设计深化，细化推敲各设计细节，并建模研究内部空间效果（集中挂图点评）。

第 5 周：制作 1：20 或 1：30 剖透视图和各分析图，制作 1：50 表现模型。

第 6 周：整理图纸、排版并完成课程答辩。

成果要求

A1 灰度图纸 2 张，纸质表现模型 1 个（1：50），场地模型 1 个（1：100），工作模型若干。图纸内容应包括：

1）总平面图（1：200），各层平面图、纵横剖面图和主要立面图（1：50），墙身大样（1：10），内部空间组织剖透视图 1 张（1：20）。

2）设计说明和主要技术经济指标（用地面积、建筑面积、容积率、建筑密度）。

3）表达设计意图和设计过程的分析图（体块生成、功能分析、流线分析、结构体系等）。

4）纸质模型照片与电脑效果图、照片拼贴等。

Course content

The main task of this exercise is to comprehensively use the knowledge points in the early case study—how to use height, opening and other operations to divide space in the horizontal direction of the building, functional streamline organization and line of sight relationship of internal space, construction mode of wall body, nodes, wrapping system and frame structure, the influence of the surrounding environment on the space, function and wrapping system, and preliminarily experience the design process of a small independent living space.

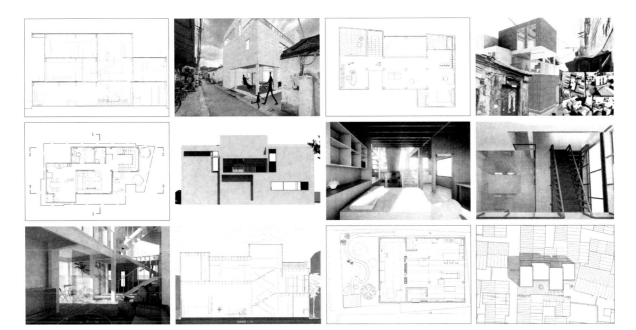

Teaching essentials

1) Site and interface: The site of this design covers an area of about 80–100 m², facing the street on one side or two adjacent sides, surrounded by traditional residential buildings of 1–2 stories.

2) Function and space: The building function of this design is a small family independent residence (with study function attached). The main members of the family include a young couple and 1–2 children (about 7 years old). The new building area is 160–200 m² and the building height is ≤ 9 m (no underground space). The designer determines the function of the space according to the set occupation and interests of family members (the occupation can be but not limited to technicians of science, engineering, medicine and law).

3) Streamline organization and entrance and exit setting: Consider the rationality of the internal streamline of the building and the reasonable connection between the entrance and exit of the building and the surrounding environmental conditions of the site.

4) Scale and perception: The size of each functional space in the building needs to take the human body scale and human behaviors as the basic reference, and express the relationship between spatial constituent elements and human spatial experience through diagrams.

Teaching progress

This design course is 6 weeks in total.

Week 1: Conceive and write several representative life scenes (composition of family characters, relative relationship of characters, a script similar to "acts" in drama). Make 1:100 (60 cm × 60 cm) site models for the four plots respectively. Use 1:100 hand-drawn drawings and the 1:100 block model to construct the internal functional spaces and their relationship.

Week 2: Draw 1:50 hand-drawn plans, elevations and sections, and use 1:50 working models to assist with design, considering function and space, streamline and scale on the basis of preliminary scheme.

Week 3: Determine the design scheme, make the spatial relationship models reflecting the functional relationship (such as three-dimensional bubble diagram), and promote the designs of section and elevation.

Week 4: Deepen the design, refine the design details, and build models to study the internal space effect (collective display and comment).

Week 5: Make 1:20 or 1:30 sectional perspective views and analysis diagrams, and make 1:50 performance models.

Week 6: Organize drawings, do the layout and complete course defense.

Achievement requirements

Two A1 grayscale drawings, one paper performance model (1:50), one site model (1:100) and several working models. The contents of the drawings shall include:

1) General plan (1:200), plans of each floor, vertical and horizontal sections and main elevation (1:50), wall detail (1:10), one sectional perspective view of internal space organization (1:20).

2) Design description and main technical and economic indicators (land area, building area, plot ratio and building density).

3) Analysis diagrams expressing design intention and design process (block generation, function analysis, streamline analysis, structural system, etc.

4) Paper model photos, computer renderings, photo collages, etc.

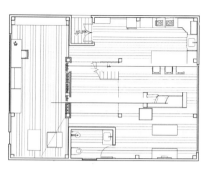

总平面图　　　　　　　　　首层平面图

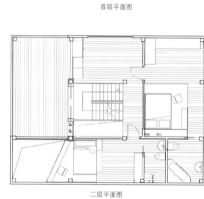

剖面图　　　　　　　　　　二层平面图

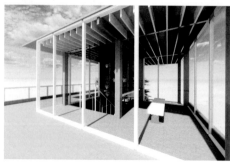
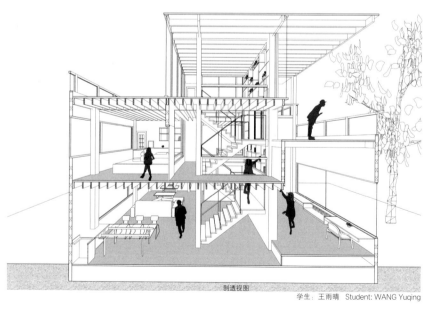

剖透视图

学生：王雨晴　Student: WANG Yuqing

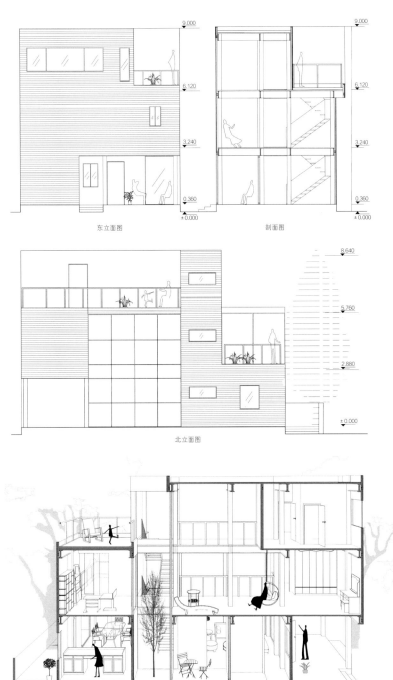
东立面图　　剖面图

北立面图

剖透视图

学生：杨雯文　Student: YANG Wenwen

校园多功能快递中心设计
CAMPUS MULTI-FUNCTIONAL EXPRESS CENTER DESIGN

冷天　何仲禹　吴佳维　孟宪川
LENG Tian　HE Zhongyu　WU Jiawei　MENG Xianchuan

课程内容

在社会信息化、电商化程度日益提高的背景下，"快递"活跃且丰富地改变了人们的日常生活，成为保障基本生活需求的重要方式。其中，高校内的快递行为富集，快递中心渐渐成了校园后勤服务中不可或缺的一环，与师生日常活动密不可分，成为校园生活区像食堂、公共浴室、超市一样重要的基础公共设施。本次练习的主要任务是在南京大学鼓楼校区南园建设一个校园多功能快递中心，要求综合运用建筑设计基础课程的知识点，操作一个小型公共建筑设计项目。

设计场地

设计训练场地位于南京大学鼓楼校区南园中轴线的西侧界面上的 110 报警中心所在地块，其南侧是南园的主要建筑教学楼，东北侧是中轴线上的圆形广场，西侧为学生宿舍区。设计范围为地块南半部分，面积大约 701.58 m²，新建建筑红线在设计范围的北侧，面积约 447.96 m²，包括原快递服务中心的一层建筑及其西侧的辅助用房。

设计要求

1）功能流线与活动

公共空间的功能——空间的使用方式——是公共建筑设计的重要内容。对于"快递中心"而言，核心功能是快递服务。功能优化方面，作为校园的基础设施，可以思考在其中加入补充的特色功能，使快递中心功能更加丰富、复合，帮助提升校园空间体验；或是在特殊的背景下，设置可变空间，增加校园内的弹性空间。基于对场地和功能的思考拟定一份设定计划书，内容包括对快递中心的定位思考、快递中心服务性功能的种类、各功能面积配比等。

与食堂、公共浴室类似，快递服务也有使用高峰时段。功能流线的组织对快递空间的使用秩序至关重要，如何合理组织快递件上架、师生取寄件等行为流线，是评价快递中心空间的重要标准之一。除此之外，还要考虑快递中心在校园中的所处位置，通过调研南园现有快递点师生寄取件情况，预设应对快递服务活动给现有校园环境带来的动线、聚集与疏散、多元化社交等系列问题。

2）形体与环境对话

场地内部和周边的现状建筑与形式特征是塑造新建筑体量的基本条件。本次设计场地是南京大学鼓楼校区南园内的真实地块。场地处于从校门口到教育超市、公共浴室的必经之路上，交通方便。作为场所环境的实体存在，建筑需要呼应周边的场地环境，充分考虑周围环境要素位置的视线关系。建筑高度 ≤ 4.8 m（檐口高度，不包括女儿墙），不超过两层（一层为主，可设置夹层）。新建建筑面积约 250—300（±10%）m²。考虑到建筑位于大学校园内，形体上一方面应具有标识性和整洁性，以丰富校园景观界面；另一方面应该符合校园气质，使建筑具有在地性。利用工作模型辅助设计，探讨建筑形体与功能之间的整合关系，思考如何利用水平、垂直构件来组织空间流线和限定功能，营造丰富的视觉和空间体验。

3）材料与构造

建筑实体部分最终限定并定义了空间，建构策略可以直接指向形体的生成以及最终效果。选择合适的材料和结构形式，利用建构的思维提出解决空间问题的策略，呼应场地需求。

Course content

Against the background of increasing social informatization and e-commerce, "express delivery" has actively and richly changed people's daily lives, and become an important way to ensure basic living needs. Among them, express delivery behaviors in colleges and universities are enriched, and express delivery center has gradually become an indispensable part of campus logistics services. It is inseparable from the daily activities of teachers and students, and has become an important basic public facility in campus living areas like canteens, public bathrooms, and supermarkets. The main task of this exercise is to build a campus multi-functional express center in the Nanyuan of Gulou Campus of Nanjing University. It is required to comprehensively use the knowledge points of the basic course of architectural design to operate a small public building design project.

Design site

The design training site is located on the west interface of the central axis of the Nanyuan of the Gulou Campus of Nanjing University, on the plot where

the Police centre is located. To its south side is the main teaching building of the Nanyuan, to the northeast is the circular square on the central axis, and on the west side is the student dormitory area. The design scope covers the southern half of the plot, with an area of about 701.58 m^2. The red line of the new building is on the north side of the design scope, with an area of about 447.96 m^2, including the one-story building of the original express center and the auxiliary buildings on the west side.

Design requirements

1) Functional flow and activities

The function of public space—the way that space is used—is an important aspect of the design of public buildings. For "Express Center", the core function is express service. In terms of function optimization, as the basic service facilities of the campus, we can consider adding supplementary features to make the functions of the express delivery center richer and more complex, and help improve the campus space experience; or in the special context, set up a variable space to increase the flexible space on campus. Based on the consideration of the site and functions, a setting plan is drawn up, which includes thinking about the positioning of the express center, the types of service functions of the express center, and the proportion of each functional area.

Similar to canteens and public bathrooms, express delivery services also have peak hours of use. The organization of functional streamlines is crucial to the use order of the express space. How to rationally organize the behavior flow such as putting express parcels on the shelves, teachers and students picking up and sending parcels, etc. is one of the important criteria for evaluating the express center space. In addition, the location of the express service center in the campus should also be considered. By investigating the situation of teachers and students sending and picking up items at the existing courier points in Nanyuan, it is necessary to presuppose a series of problems brought about by express service activities to the existing campus environment, such as gathering and evacuation, diversified social interaction and other issues.

2) Interaction between form and environment

The existing architecture and form characteristics in side and around the site are the basic conditions for shaping the volume of the new building. The site for this design is a real plot in the Nanyuan of the Gulou Campus of Nanjing University. The site is located on the only way from the school gate to the education supermarket and the public bathroom, with convenient transportation. As the physical existence of the site environment, the building needs to respond to the surrounding site environment and fully consider the sight relationship between the positions of the surrounding environmental elements. The building height should be no more than 4.8 m (the height of the eaves, excluding the parapet wall), should have no more than two floors (mainly one floor, interlayer can be set). The new construction area is about 250–300 (±10%) m^2. Considering that the building is located on the university campus, on the one hand, its shape should be marked and tidy to enrich the campus landscape interface; on the other hand, it should conform to the campus temperament, making the building local. Use the working model to assist design, explore the integration relationship between architectural form and function, think about how to use horizontal and vertical components to organize space streamlines and limit functions, and create a rich visual and spatial experience.

3) Materials and construction

The physical part of the building finally defines the space, and the construction strategy can directly point to the generation of the shape and the final effect. Choose appropriate materials and structural forms, use constructive thinking to propose strategies to solve space problems, and respond to site needs.

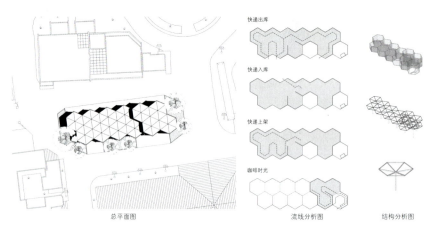

总平面图　　　　　　　　　流线分析图　　　结构分析图

快递出库
快递入库
快递上架
咖啡时光

一层平面图

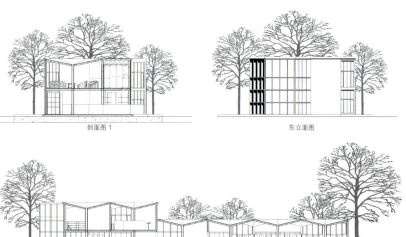

剖面图1　　　　　　　东立面图

剖面图2

学生：申翱　Student: SHEN Ao

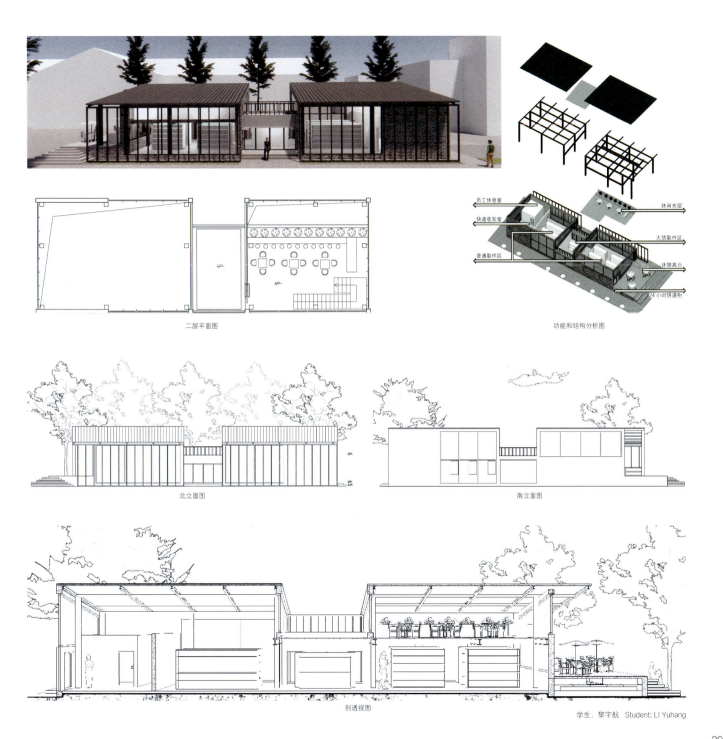

建筑与规划设计（二）ARCHITECTURE AND PLANNING DESIGN 2

旧城中的新社区
NEW COMMUNITY IN OLD TOWN

童滋雨　窦平平　张益峰
TONG Ziyu DOU Pingping ZHANG Yifeng

教学目标

从空间单元到系统组合。

从个体到整体，从单元到体系，是建筑空间组织的一种基本和常用方式。基本单元的重复、韵律、节奏、变异等都是常用的操作手法。关注空间单元的生成，同时充分考虑人在单元重复的空间中的行为和感受。将多个单元通过特定方式与秩序组合起来，形成一个兼具合理性、清晰性和丰富性的整体系统。

构想低层紧凑、建筑与景观密切结合的居住模式。

在空间单元和单元组合的设计过程中，将景观要素与建筑要素同时设计，利用景观要素增进高密度城市住宅的居住体验，满足低层紧凑的空间需求，实现尊重旧城肌理的理想新社区。

基本任务

基地位于南京市门西荷花塘历史片区，东接刘芝田故居，西邻鸣羊街，北靠殷高巷，南侧为鸣羊里，用地面积约 3 100 m^2。拟建共享公寓，总建筑面积约 3 000 m^2，高度不超过 12 m。场地周边历史遗存丰富，但物质空间衰败严重，原居民流失，城市功能相对滞后。

基本面积需求：

1) 公共活动与对外营业空间（约 900 m^2）

门厅和服务台：100 m^2；

咖啡简餐厅：80 m^2 × 4 间；

展览空间（可结合走廊、过厅）：150 m^2；

文创商店：80 m^2；

多功能空间：100 m^2 × 2 间。

2) 共享公寓（约 2 000 m^2）

单人公寓房间：（30—40）m^2 × 24 间；

双人公寓房间：（60—80）m^2 × 10 间；

共享客厅和餐厨房：80 m^2；

共享健身房：100 m^2；

创意办公空间：（30—40）m^2 × 4 间。

3) 辅助空间（约 200 m^2）

管理办公室：15 m^2 × 2 间；

服务间：15 m^2 × 2 间；

公共卫生间、储藏室等：按需配置。

教学进度

第 1 周：布置设计任务书，现场踏勘或结合地图和资料熟悉场地，搜集与分析国内外案例。

第 2 周：分析任务书，学习单元组合的设计方法，提出设计思路。

第 3 周：给出概念方案与体量策略。

第 4 周：深化平面布局。

第 5 周：深化单元设计。

第 6 周：重点空间设计与细部设计。

第 7 周：完成设计，深化图纸。

第 8 周：深化图纸，准备答辩。

Teaching objectives

From spatial units to system assemblies.

From the individual to the whole, from units to systems, it is a basic and common way of architectural space organization. Repetition, rhythm, tempo and variation of basic units are commonly used manipulation techniques. Pay attention

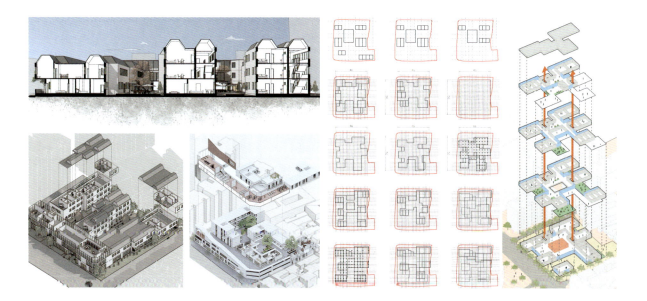

to the generation of spatial units, and at the same time fully consider people's behaviors and feelings in the space where units are repeated. Combines multiple units with order in a specific way to form an overall system with rationality, clarity and richness.

Conceive a residential pattern that is low-rise and compact, with a close integration of architecture and landscape.

In the design process of spatial units and unit combinations, landscape elements and architectural elements are designed at the same time, and landscape elements are used to enhance the living experience of high-density urban housing, meet the spatial demands of low-rise and compact layouts, and realize the ideal new community that respects the fabric of the old town.

Basic tasks

The site is located in the Hehuatang historical block of Menxi area, Nanjing, with Liu Zhitian's former residence to the east, Mingyang Street to the west, Yingao Lane to the north and Mingyang Alley to the south, covers an area of about 3,100 m^2. It is planned to build a shared apartment with a total building area of about 3,000 m^2 and a height of no more than 12 m. There are abundant historical relics around the site, but the physical space is seriously decayed, the original residents have been lost, and the urban functions are relatively lagging behind.

Basic area requirements:
1) Public activities and business space open to the outside (about 900 m^2)
Foyer and reception desk: 100 m^2;
Cafe and Bistro: 80 m^2 × 4;
Exhibition space (which can be combined with corridors and halls): 150 m^2;
Cultural and creative store: 80 m^2;
Multi-functional spaces: 100 m^2 × 2.
2) Shared apartments (about 2,000 m^2)
Single apartment rooms: (30–40) m^2 × 24;
Double apartment rooms: (60–80) m^2 × 10;
Shared living room and dining kitchen: 80 m^2;
Shared gym: 100 m^2;
Creative office spaces: (30–40) m^2 × 4.
3) Auxiliary spaces (about 200 m^2)
Management offices: 15 m^2 × 2;
Service rooms: 15 m^2 × 2;
Public toilets, storage rooms, etc.: Set according to needs.

Teaching schedule

Week 1: Arrange the design assignment, conduct on-site surveys or get familiar with the site according to maps and materials, and collect and analyze domestic and overseas cases.

Week 2: Analyze the task, study the design method of unit combination, and put forward the design idea.

Week 3: Propose conceptual schemes and volume strategies.
Week 4: Deepen the planar layout.
Week 5: Deepen the unit design.
Week 6: Design key space and details.
Week 7: Complete the design and deepen the drawings.
Week 8: Deepen the drawings, and prepare for the presentation.

A-A 剖面图

北立面图

B-B 剖面图

西立面图

学生：王雨晴 Student: WANG Yuqing

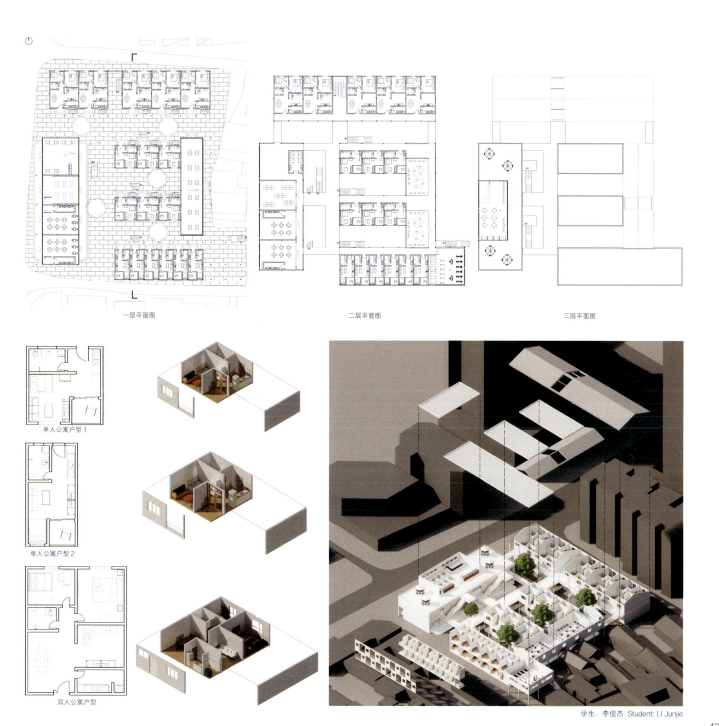

都市田园共享公寓
URBAN GARDEN SHARED APARTMENT

窦平平
DOU Pingping

教学目标

从空间单元到系统组合。

从个体到整体,从单元到体系,是建筑空间组织的一种基本和常用方式。基本单元的重复、韵律、节奏、变异等都是常用的操作手法。关注空间单元的生成,同时充分考虑人在单元重复的空间中的行为和感受。将多个单元通过特定方式与秩序组合起来,形成一个具备合理性、清晰性和丰富性的整体系统。

构想低层紧凑、建筑与景观密切结合的居住模式。

在空间单元和单元组合的设计过程中,将景观要素与建筑要素同时设计,利用景观要素增进高密度城市住宅的居住体验,满足低层紧凑的空间需求,实现都市田园的理想公寓模式。

基本任务

基地位于南京市颐和路历史保护区03片区,颐和路2号,用地面积1 750 m²。场地周边为民国时期的花园住宅,拟新建适应当代需求的都市田园共享公寓,总建筑面积约3 200 m²,高度不超过24 m。

基本面积需求:

1)公共活动与对外营业空间(约900 m²)

门厅和服务台:100 m²;

咖啡简餐厅:80 m²×4间;

展览空间(可结合走廊、过厅):150 m²;

文创商店:80 m²;

多功能空间:100 m²×2间。

2)共享公寓(约2 000 m²)

单人公寓房间:(30—40)m²×24间;

双人公寓房间:(60—80)m²×10间;

共享客厅和餐厨房:80 m²;

共享健身房:100 m²;

创意办公空间:(30—40)m²×4间。

3)辅助空间(约200 m²)

管理办公室:15 m²×2间;

服务间:15 m²×2间;

公共卫生间、储藏室等:按需配置。

教学进度

第1周:布置设计任务书,现场踏勘或结合地图和资料熟悉场地,搜集与分析国内外案例。

第2周:分析任务书,学习单元组合的设计方法,提出设计思路。

第3周:给出概念方案与体量策略。

第4周:深化平面布局。

第5周:深化单元设计。

第6周:重点空间设计与细部设计。

第7周:完成设计,深化图纸。

第8周:深化图纸,准备答辩。

Teaching objectives

From spatial units to system assemblies.

From the individual to the whole, from units to systems, it is a basic and common way of architectural space organization. Repetition, rhythm, tempo and variation of basic units are commonly used manipulation techniques. Pay attention to the generation of spatial units, and at the same time fully consider people's

behaviors and feelings in the space where units are repeated. Combines multiple units with order in a specific way to form an overall system with rationality, clarity and richness.

Conceive a residential pattern that is low-rise and compact, with a close integration of architecture and landscape.

In the design process of spatial units and unit combinations, landscape elements and architectural elements are designed at the same time, and landscape elements are used to enhance the living experience of high-density urban housing, meet the spatial demands of low-rise and compact layouts, and realize the ideal new community that respects the fabric of the old town, and realize the ideal apartment mode of an urban garden.

Basic tasks

The site is located in Area 03 of the Yihe Road Historical Reserve in Nanjing, at No. 2 Yihe Road, with a land area of 1,750 m^2. The site is surrounded by garden houses from the Republic of China period, and it is planned to build a new urban garden shared apartment to meet contemporary needs, with a total building area of about 3,200 m^2 and a height of no more than 24 m.

Basic area requirements:
1) Public activities and business space open to the outside (about 900 m^2):
Foyer and reception desk: 100 m^2;
Cafe and Bistro: 80 m^2 × 4;
Exhibition space (which can be combined with corridors and halls): 150 m^2;
Cultural and creative store: 80 m^2;
Multi-functional spaces: 100 m^2 × 2.

2) Shared apartments (about 2,000 m^2)
Single apartment rooms: (30–40) m^2 × 24;
Double apartment rooms: (60–80) m^2 × 10;
Shared living room and dining kitchen: 80 m^2;
Shared gym: 100 m^2;
Creative office spaces: (30–40) m^2 × 4.
3) Auxiliary spaces (about 200 m^2)
Management offices: 15 m^2 × 2;
Service rooms: 15 m^2 × 2;
Public toilets, storage rooms, etc.: Set according to needs.

Teaching schedule

Week 1: Arrange the design assignment, conduct on-site surveys or get familiar with the site according to maps and materials, and collect and analyze domestic and overseas cases.

Week 2: Analyze the task, study the design method of unit combination, and put forward the design idea.

Week 3: Propose conceptual schemes and volume strategies.
Week 4: Deepen the planar layout.
Week 5: Deepen the unit design.
Week 6: Design key space and details.
Week 7: Complete the design and deepen the drawings.
Week 8: Deepen the drawings, and prepare for the presentation.

北立面图

东北立面图

一层平面图

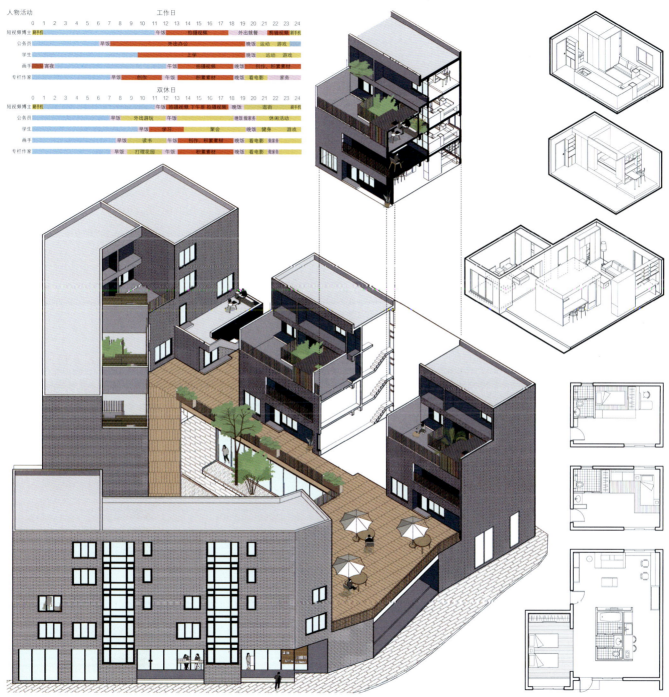

基于算法的专家公寓设计
THE EXPERT APARTMENT DESIGN BASED ON ALGORITHM

童滋雨
TONG Ziyu

教学目标

从空间单元到系统；从规则到算法。

从个体到整体，从单元到体系，是建筑空间组织的一种基本和常用方式。本课题首先关注空间单元的生成，并进一步根据内在的使用逻辑和外在的场地条件，将多个单元通过特定方式与秩序组合起来，形成一个兼具合理性、清晰性和丰富性的整体系统。基本单元的重复、韵律、变异等都是常用的操作手法。

另外，如果将设计视作对特定问题的解决方案，从场地环境、功能要求以及空间单元组织的方式出发，可以将设计转化为一系列的规则限定，进而将这些规则转化为算法。我们可以在计算机的辅助下探索能够解决问题的更多可能。在基于算法的计算性设计过程中，形式只是计算的结果，算法成为我们更加关注的对象。

基本任务

拟在南京大学鼓楼校区南园宿舍区内新建专家公寓一座，用于国内外专家到访南京大学开展学术交流活动期间的居住。用地位于南园中心喷泉西侧，面积约 3 600 m²。地块上原有建筑将被拆除。基本规划指标：容积率 1.0，建筑密度不超过 40%，建筑高度不超过 15 m。具体的功能空间包括：

客房：36 间左右，分为单间和套间两类，单间面积约 35—40 m²，套间面积约 70—80 m²，内部需包括睡眠空间、卫生间、学习工作空间。套间可考虑必要的接待空间和简单的餐厨空间，不少于 6 间。

大会议室：1 间，100—120 m²。

研讨室：3 间，每间约 60 m²。

休闲区与咖啡吧：兼做早餐厅，约 150 m²。

操作间：约 30 m²。

服务间：每层 1 间，每间约 20 m²。

工作人员办公室与休息室：1—2 间，每间约 30 m²。

其他必要的门厅、前台、公共卫生间、储藏室、服务间等：自行设置。

场地环境：需结合建筑总体布局，在建筑周边及其内部创造优美的室内外场地环境，供使用者休憩交往，并为校园增色。

教学成果

图纸：总平面图（1:500），建筑平、立、剖面图（1:200），客房单元平面图（1:50），分析图，轴测图，鸟瞰图，剖透视图，人眼透视图，其他有助于表达方案的图纸。

教学进度

第 1 周：布置设计任务书，根据卫星地图熟悉场地，进行相关案例收集和分析。

第 2 周：分析任务书，解析其中最关注的问题，提出解决问题的思路。

第 3 周：基于规则建立解决问题的算法，通过计算性操作完成初步的整体布局和平面组织。

第 4 周：调整算法，深化建筑平面布局。

第 5 周：继续调整算法，深化建筑平面布局。

第 6 周：在基本确定的建筑平面布局基础上完善建筑设计，包括结构、立面等。

第 7 周：深化并完成所有设计成果。

第 8 周：深化建筑图纸，完成最终设计答辩。

Teaching objectives

Design training from space unit to system and from rules to algorithms.

From individual to whole, from unit to system, it is a basic and common way of architectural space organization. This topic first pays attention to the generation of spatial units, and further combines multiple units with order in a specific way according to the internal use logic and external site conditions to form an overall system with rationality, clarity and richness. Repetition, rhythm and variation of basic units are commonly used.

Additionally, if the design is regarded as a solution to specific problems, starting from the site environment, functional requirements and the way of spatial unit organization, the design can be transformed into a series of rules

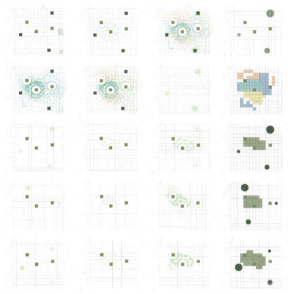

and constraints, and then these rules can be converted into algorithms. We can explore more possibilities to solve the problem with the aid of computers. In the process of algorithm-based computational design, the form is only the result of the calculation, and the algorithm becomes the object we pay more attention to.

Basic tasks

It is proposed to build a new expert apartment in the Nanyuan dormitory area of Gulou Campus of Nanjing University for domestic and foreign experts to live during their visit to Nanjing University for academic exchange activities. The site is located in the west of the fountain in the center of Nanyuan, covering an area of about 3,600 m^2. The original buildings on the plot will be demolished. Basic planning indexes: plot ratio 1.0, building density less than 40%, building height not exceed 15 m. The specific functional spaces include:

Guest rooms: About 36 rooms, divided into single rooms and suites, with a single room area of 35–40 m^2 and a suite area of 70–80 m^2. The interior needs to include sleeping space, a toilet, and a study and working space. The necessary reception space and simple kitchen space can be considered in the suite. No less than 6 suites.

One large conference room: 100–120 m^2.

Three seminar rooms: Each about 60 m^2.

Leisure area and coffee bar: As a breakfast room, about 150 m^2.

Operation room: About 30 m^2.

Service room: One room on each floor, each about 20 m^2.

Staff offices and lounges: 1–2 rooms, each about 30 m^2.

Other necessary foyers, front desks, public toilets, storage rooms, service rooms, etc.: Set independently.

Site environment: It is necessary to combine with the overall layout of the building, create a beautiful indoor and outdoor site environment around and inside the building for users to rest and communicate, and enhance the landscape of the campus.

Teaching achievements

Drawings: General layout (1 : 500), building plans, elevations and sections (1 : 200), guest room unit plan (1 : 50), analysis drawings, axonometric drawings, aerial views, sectional perspective views, human-eye perspective views, and other drawings which are helpful to express the scheme.

Teaching progress

Week 1: Arrange the design assignment, get familiar with the site according to the satellite map, and collect and analyze relevant cases.

Week 2: Analyze the assignment, identify the most concerned problems, and put forward the ideas to solve the problems.

Week 3: Build a rule-based and problem-solving algorithm and complete the initial overall layout and plan organization through computational operations.

Week 4: Adjust the algorithm to deepen the the plane layout.

Week 5: Continue to adjust the algorithm to deepen the plane layout.

Week 6: Improve the architectural design on the basis of the basically determined plane layout, including structure, facade, etc.

Week 7: Deepen and complete all design results.

Week 8: Deepen the architectural drawings and complete the final presentation.

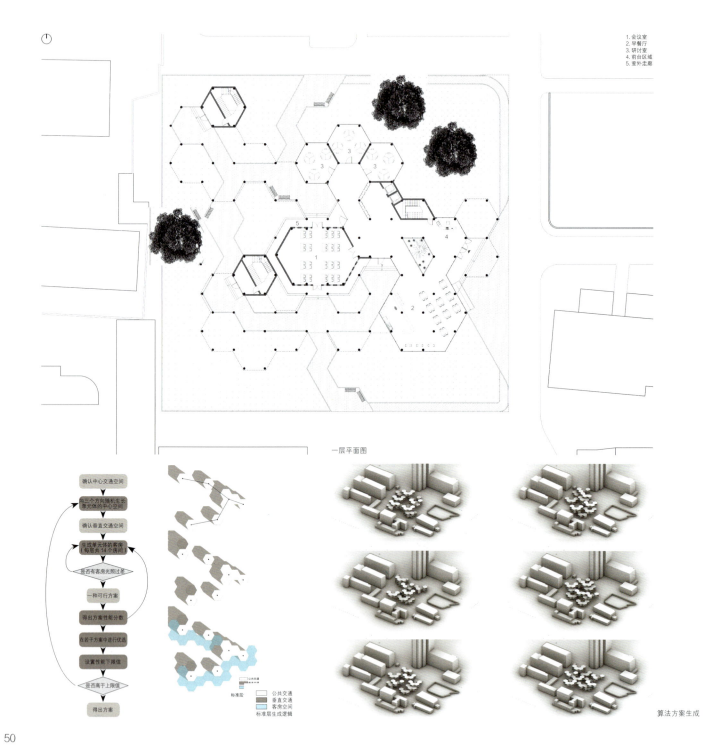

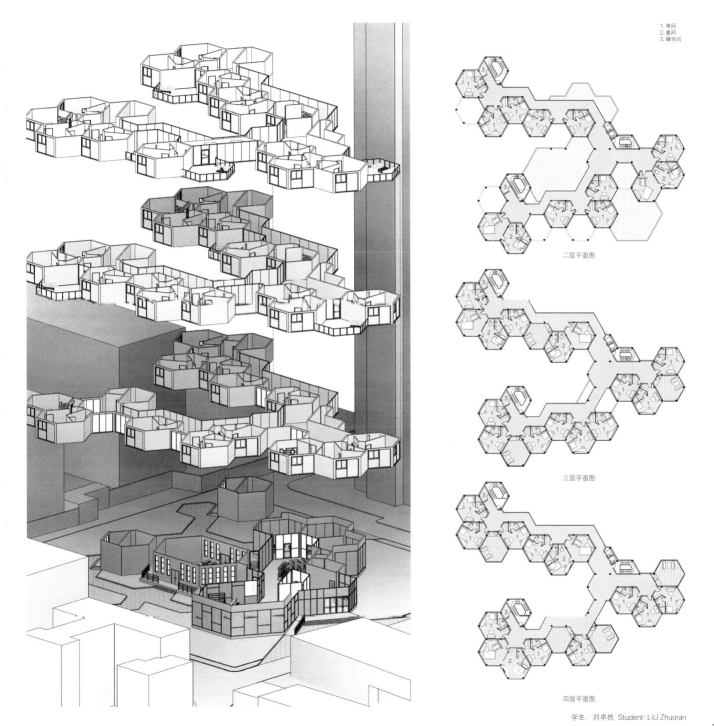

1. 单间
2. 套间
3. 操作间

二层平面图

三层平面图

四层平面图

学生：刘卓然　Student: LiU Zhuoran

集装箱中学设计
THE DESIGN OF CONTAINER MIDDLE SCHOOL

钟华颖
ZHONG Huaying

教学目标
从标准单元到个性化校园。

单元式建筑的优势在于标准化设计带来的效率，特别是加工建造缩减了建设成本和时间周期。另外，单元的标准化对设计形成制约。设计的灵活性、场地的适应性等限制条件对单元式建筑的设计提出针对性研究需求。如何在保证单元式建筑便捷性的同时形成个性化的设计，提升单元式建筑的体验感与建成环境品质，是此类型建筑设计需要重点考虑的因素。

基本任务
为在新校区建设期间临时安置在读学生而设计一所使用期三年、由集装箱构成的临时学校——江心洲集装箱中学。临时学校与用地南侧小学合用运动场地，家长接送临时停车利用南侧城市道路。建筑及场地景观构件材料尽可能可以回收利用。树木绿化种植考虑后期可移栽。

现状基地为一块下凹空地，基地面积为 11 221.4 m²，基地南侧为现状小学。拟在基地内建设教学楼、行政楼及门卫各一栋。包含以下功能：

1) 教学楼（建筑面积约 1 700 m²）
普通教室：60 m² × 9 间；
音乐教室：60 m²；
计算机教室：70 m²；
图书阅览室：70 m²；
教师办公室：60 m² × 3 间；
其他辅助空间：卫生间、开水间等。

2) 行政楼（建筑面积约 1 050 m²）
行政办公室：若干，合计面积约 350 m²；
文印室：35 m²；
广播室：18 m²；
体育器材室：35 m²；
卫生保健室：35 m²；
会议室：50 m²；
其他辅助空间：卫生间、开水间等。

3) 门卫（建筑面积约 35 m²）
门卫室：兼消防安防控制室，设置在场地出口。
注：以上均为使用面积，不含交通面积。

教学成果
不少于 4 张 A1 图纸，图纸内容包括：
1) 建筑与环境：1∶500 总平面图。
2) 建筑基本表达：1∶200 平、立、剖面图。
3) 建筑解析与表现：形体概念生成分析图、结构单元生成分析图、剖透视图、鸟瞰图不少于 1 张，室外、室内人视点透视图若干。
4) 构造方式与表达：1∶50 构造细部图。

教学进度
第 1 周：布置设计任务书，现场踏勘或结合地图和资料熟悉场地，搜集与分析国内外案例。
第 2 周：分析任务书，学习单元组合的设计方法，提出设计思路。
第 3 周：给出概念方案与体量策略。
第 4 周：深化平面布局。
第 5 周：深化单元设计。
第 6 周：重点空间设计与细部设计。
第 7 周：完成设计，深化图纸。
第 8 周：深化图纸，准备答辩。

Teaching objectives
From standard units to a personalized campus.

The advantage of the unit building is the efficiency that the standardized design brings, especially the reduction in construction costs and time cycles due to the processing and construction. Additionally, the standardization of the unit restricts the

design, so the flexibility of the design, the adaptability of the site and other restrictions requires to carry out targeted research on the design of the unit building. How to ensure the convenience of the unit building while forming a personalized design, improving the experience of the unit building and the quality of the built environment, is the key factor to be considered in the design of this type of architecture.

Basic tasks

In order to temporarily relocate students during the construction of the new campus, Jiangxinzhou Container Middle School, is designed as a temporary school composed of containers. The school has a service life of three years. The temporary school shares sports field with the primary school on the south side of the site, and uses the urban road on the south side for temporary parking during pick-up and drop-off. Materials for building and landscape components should be recyclable as much as possible. Trees and green plants can be considered to transplant in the later stage.

The current base is a concave space with an area of 11,221.4 m^2. The south of the base is the current primary school. It is planned to build one teaching building, one administrative building and one guard room in the site. The functions include:

1) Teaching building (building area about 1,700 m^2)
Ordinary classrooms: 60 m^2 × 9;
Music classroom: 60 m^2;
Computer classroom: 70 m^2;
Reading room: 70 m^2;
Teacher's offices: 60 m^2 × 3;
Other auxiliary spaces: Toilet, water heater room, etc.
2) Administration building (building area about 1,050 m^2)
Administrative offices: 350 m^2 in total;
Printing room: 35 m^2;
Broadcasting room: 18 m^2;
Sports equipment room: 35 m^2;
Medical room: 35 m^2;
Meeting Room: 50 m^2;
Other auxiliary spaces: Toilet, water heater room, etc.
3) Guard room (building area about 35 m^2)
Guard room: Serve as the fire security control room, to be set at the site exit.
Note: The above areas are all usable areas, excluding traffic areas.

Teaching achievements

No less than 4 A1 drawings, and the drawings include:
1) Architecture and environment: 1∶500 general layout.
2) Basic architectural expression: 1∶200 plans, elevations and sections.
3) Architectural analysis and performance: Analysis drawings of form concept generation, analysis drawings of structural unit generation, sectional perspective views and aerial views, each no less than one, and several perspective views door and outdoor human-eye perspective.
4) Construction and expression: 1∶50 construction detail drawings.

Teaching schedule

Week 1: Arrange the design assignment, conduct on-site surveys or get familiar with the site according to maps and materials, and collect and analyze domestic and overseas cases.
Week 2: Analyze the task, study the design method of unit combination, and put forward the design idea.
Week 3: Propose conceptual schemes and volume strategies.
Week 4: Deepen the planar layout.
Week 5: Deepen the unit design.
Week 6: Design key space and details.
Week 7: Complete the design, and deepen the drawings.
Week 8: Deepen the drawings, and prepare for the presentation.

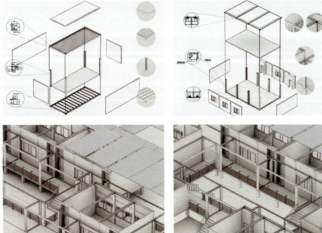

1. 教室
2. 办公室
3. 公共教室
4. 行政
5. 卫生间

层平面图

学生：李若松 Student: LI Ruosong

建筑设计（四）ARCHITECTURAL DESIGN 4
世界文学客厅
WORLD LITERATURE LIVING ROOM

华晓宁　尹航　梁宇舒
HUA Xiaoning　YIN Hang　LIANG Yushu

教学目标

本课程主题是"空间"，学习建筑空间组织的技巧和方法，训练对空间的操作与表达。空间问题是建筑学的基本问题。本课题基于文学主题，训练文本、叙事与空间序列的串联，学习空间叙事与空间用途的整体构思，充分考虑人在空间中的行为、空间感受，尝试以空间为手段表达特定的意义和氛围，最终形成一个完整的设计。

课程内容

南京古称金陵、白下、建康、建邺……历来是人文荟萃、名家辈出之地，号称"天下文枢"。南京作为六朝古都，亦为中国文学之始。何为文？梁元帝曰："吟咏风谣，流连哀思者，谓之文。"汉魏有文无学，六朝文学《文选》《文心雕龙》《诗品》既是文学评论的开始，也是文学的发端。

2019年，南京入选联合国"世界文学之都"，开展一系列城市空间计划，包括筹建"世界文学客厅"，作为一座以文学为主题的综合性博物馆。该馆计划位于台城花园与台城城墙之间，解放门西侧，用地面积约6 500 m²，毗邻古鸡鸣寺、玄武湖、明城墙等历史文化遗迹，构成城市与山林之间的过渡空间。设计应妥善处理建筑与周边城市环境和既有建筑的关系，彰显中国文学的精神特质。

空间计划

建筑地面以上原则上不超过2层，退让城墙不少于10 m，绿地率不低于25%

空间组织要有明确的特征和意图，概念清晰；满足功能合理、环境协调、流线便捷的要求。注意不同类型和不同形态空间的构成、串联组织和氛围的塑造。

总建筑面积：约3 000 m²。

1）文学之都展示中心

主题展厅：700—800 m²；

临时展厅：100—150 m²；

其他辅助空间：控制室、储藏间等。

2）文学交流中心

报告厅：100座，150—200 m²；

会议室：4—6间，共300 m²；

会客厅：60—80 m²；

其他辅助空间：休闲、接待等。

3）城市客厅

游客服务：150—200 m²。

4）城市书房

书籍展示及阅览：150—200 m²。

5）行政办公与辅助空间

办公室：6间，每间不小于15 m²；

专家接待（客房）：4间，每间不小于35 m²；

保安值班室：30 m²，有单独对外出口；

其他门厅、交通、设备间、卫生间等：面积根据设计需要自行确定。

6）场地

园林景观、户外展场、停车场（内部工作人员使用）等：应考虑建筑与景观的整体关系，以景观烘托氛围。

Teaching objectives

The theme of this course is "space", learning the skills and methods of architectural space organization, and training the operation and expression of space. The space problem is the basic problem of architecture. Based on the literary theme, this topic trains the series of the text, narration and spatial sequence, learns the overall idea of spatial narration and spatial use, fully considers people's behaviors and spatial feelings in space, tries to express specific meaning and atmosphere by means of space, and finally forms a complete design.

Course content

In ancient times, Nanjing was called Jinling, Baixia, Jiankang, Jianye...It has

always been a place with a large number of talents and famous scholars, known as the "Cultural Hub of the World". As the ancient capital for Six Dynasties, Nanjing is also the beginning of Chinese literature. What is literature? Emperor Liang Yuandi once said, "It is called literature to chant wind ballads and linger around mourning." The Han and Wei Dynasties have literature but no learning. The literature of the Six Dynasties *Selections of Refined Literature*, *The Literary Mind and the Carving of Dragons*, and *The Critique of Poetry* are not only the beginning of literary criticism, but also the beginning of literature.

In 2019, Nanjing was selected as a "City of Literature" by the United Nations and planned to carry out a series of urban space plans, including the preparation for the construction of the "World Literature Living Room" as a comprehensive museum with literature as the theme. The museum is planned to be located between Taicheng Garden and Taicheng City Wall, on the west side of Jiefang Gate, covers an area of about 6,500 m^2, adjacent to ancient Jiming Temple, Xuanwu Lake, Ming City Wall and other historical and cultural relics, forming a transitional space between the city and the mountains. The design should deal with the relationship between the building and the surrounding urban environment and existing buildings properly, and highlight the spiritual characteristics of Chinese literature.

Space program

In principle, the building should not exceed 2 floors above the ground, and the setback distance from city walls should not be less than 10 m. The green space rate shall not be less than 25%.

Spatial organization should have clear characteristics, clear intentions and clear concepts, to meet the requirements of reasonable function, coordinated environment and convenient streamline. Pay attention to the composition of different types and forms of space, the series organization of space as well as the shaping of space atmosphere.

Total construction area: About 3,000 m^2.
1) Exhibition center of the City of Literature
Theme exhibition hall: 700–800 m^2;
Temporary exhibition hall: 100–150 m^2;
Other auxiliary spaces: Control rooms, storage rooms, etc.
2) Literature communication center
Lecture hall: 100 seats, 150–200 m^2;
Meeting rooms: 4–6 rooms, 300 m^2 in total;
Reception room: 60–80 m^2 in total;
Other auxiliary spaces: Leisure, reception, etc.
3) City living room
Tourist service: 150–200 m^2.
4) City library
Book display and reading: 150–200 m^2.
5) Administrative office and auxiliary space
Offices: 6 rooms, each not less than 15 m^2;
Expert reception (guest rooms): 4 rooms, each not less than 35 m^2;
Security duty room: 30 m^2, with an independent external exit;
Foyers, traffic spaces, equipment rooms, toilets and other spaces: The area shall be determined according to the design needs.
6) Site
Garden landscape, outdoor exhibition halls and parking lots (for internal staffs): Consider the overall relationship between architecture and landscapes, so as to set off the atmosphere with landscapes.

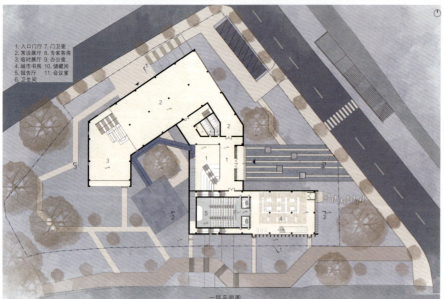

1. 入口门厅 7. 门卫室
2. 常设展厅 8. 专家客房
3. 临时展厅 9. 办公室
4. 城市书房 10. 储藏间
5. 报告厅 11. 会议室
6. 卫生间

一层平面图

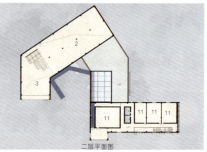

二层平面图

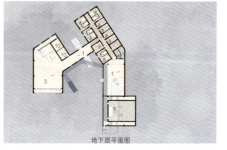

地下层平面图

1-1 剖面图　　　　　　　　　　　　　　　2-2 剖面图

南立面图　　　　　　　　　　　　　　　3-3 剖面图

学生：李静怡　Student: LI Jingyi

大学生健身中心改扩建设计
RECONSTRUCTION AND EXPANSION DESIGN OF COLLEGE STUDENT FITNESS CENTER

傅筱 钟华颖 孟宪川
FU Xiao ZHONG Huaying MENG Xianchuan

教学目标

本项目拟在南京大学鼓楼校区体育馆基地处改扩建大学生健身中心，以服务于南京大学师生，可适当考虑对周边居民的服务。根据基地条件、功能要求进行建筑和场地设计。

本课题以大学生健身中心为训练载体，在学习中小跨建筑的基本设计原理的基础上，理解建筑形式、空间与基本的结构类型与之间的逻辑关系，初步培养学生整合建筑内外空间、建筑结构、建筑场地，以及兼顾合理的建筑采光通风节能的综合能力。

教学内容

现状基地由一座体育馆和一座游泳馆（吕志和馆）组成，设计需保留体育馆，拆除现状游泳馆并重新设计一座大学生健身中心，其建筑总面积约4 300 m²（可上下浮动10%）。建筑高度控制在24 m以下，注意场地东西向高差，场地下挖不得超过一层，深度不超过4.5 m。主要包含以下功能：

1）公共活动
（1）游泳馆（使用面积约1 650 m²）
25 m游泳池：一片，池面25 m×18.5 m—25 m×21 m（6—8个泳道）；
儿童戏水池：平面形状不限，150 m²左右；
更衣、淋浴、卫生间：120 m²×2间；
公共卫生间：18 m²×2间；
体检室：18 m²；
急救室：18 m²；
救生员休息室：18 m²；
设备控制室：18 m²。
（2）健身中心（使用面积约680 m²）
健身房：180 m²；
体操厅：180 m²；
瑜伽室：120 m²；
普拉提教室：120 m²；
更衣、淋浴：18 m²×2间。
2）辅助空间
管理办公用房：18 m²×3间；
器材库房：60 m²；
消防控制室兼监控：30 m²，一楼，有独立对外出口；
强、弱电间：各6 m²，每层均有，上下对位；

新风机房：18 m²，每层均有；
空调机房：60 m²；
水处理机房：100 m²，地下，应靠近泳池；
配电房：100 m²，地下；
消防水池：120 m²，地下；
消防水泵房：60 m²，地下；
门厅、体育用品专卖、咖啡吧、交通等公共部分：自行策划。
3）室外场地
保留场地原有大树，场地设计满足15辆机动车和200辆非机动车停车要求，在条件允许时预留一定的室外活动场地。

设计要求

1）老羽毛球馆建筑主体结构不宜改动，外观立面根据需要可进行适度改造以保持改扩建风格的协调性。
2）通过内外公共空间的塑造，激活场地活力，并充分考虑与场地周边空间尺度的协调性。
3）结构选型：每四个同学为一组，在木、钢、混凝土三种材料中选择一种或两种材料相结合的结构类型进行中小跨空间训练，也可以根据方案进展的需求选择合适的结构类型。
4）覆面材料选择：注意屋顶材料与结构选型的相互配合。注意游泳大厅外围护材料的保温性能与防起雾功能。
5）空调方式：游泳馆大厅及戏水池采用集中空调和机械通风。其余用房采用变制冷剂流量（VRV）空调，使用功能用房均应考虑春秋季及过渡季节采用自然采光和通风。
6）鼓励采用Revit推进课程设计。

Teaching objectives

This project intends to rebuild and expand the college student fitness center at the gymnasium base of the Gulou Campus of Nanjing University to serve the teachers and students of Nanjing University, and give due consideration to the service of surrounding residents. Architecture and site design are based on the base conditions and functions.

This project takes the college student fitness center as a training platform.

By learning the fundamental design principles of small and medium-size span buildings, it aims to help students comprehend the logical relationship between architectural forms, space and basic structural types. Additionally, it aims to cultivate students' comprehensive ability in integrating architectural internal and external spaces, building structures and sites, as well as ensuring reasonable building lighting, ventilation, and energy conservation.

Teaching content

The current site comprises a gymnasium and a swimming pool (Lui Che-woo natatorium). The design requires that the gymnasium should be retained and the existing swimming pool should be demolished. A new fitness center for college students needs to be redesigned with a total area of approximately 4,300 m^2 (10% allowance). The height of the building should not exceed 24 m while considering the east-west height difference of the site. Excavation under the site should be limited to one level with a maximum depth of 4.5 m. The main functions include:

1) Public activities

(1) Swimming pool (approximately 1,650 m^2)

25 meter swimming pool: One pool with a length of 25 m ranging from 18.5 to 21 m in width (6–8 lanes);

Children's paddling pool: No specific shape requirement, covering about 150 m^2;

Dressing rooms, showers, toilets: 120 m^2 × 2;

Public toilets: 18 m^2 × 2;

Medical examination room: 18 m^2;

Emergency room: 18 m^2;

Lifeguard Lounge 18 m^2;

Equipment control room: 18 m^2.

(2) Fitness center area (usable area of about 680 m^2)

Gymnasium: 180 m^2;

Gymnastics hall: 180 m^2;

Yoga room: 120 m^2;

Pilates room: 120 m^2;

Dressing rooms and showers: 18 m^2 × 2.

2) Auxiliary spaces

Management offices: 18 m^2 × 3;

Equipment warehouse: 60 m^2;

Fire control and monitoring room: 30 m^2, on the first floor, with an independent exit;

Strong and weak current rooms: Each for 6 m^2, on every floor, up and down matching;

Fresh air handling rooms: 18 m^2, on every floor;

Air conditioning room: 60 m^2;

Water treatment room: 100 m^2, underground, should be near the swimming pool;

Power distribution room: 100 m^2, underground;

Fire water tank: 120 m^2, underground;

Fire pump room: 60 m^2, underground;

Public areas such as foyers, sports goods stores, coffee bars, and connection space: Arranged optionally.

3) Exterior site

The original trees on the site should be retained, the site design should meet the parking requirements of 15 motor vehicles and 200 non-motor vehicles, and some outdoor activity areas should be reserved when conditions permit.

Design requirements

1) The main structure of the old badminton hall should not be changed, and the facade can be moderately transformed according to needs to maintain the coordination of the renovation and expansion style.

2) Activate the vitality of the site through shaping the internal and external public spaces, and fully consider the coordination with the scale of surrounding space.

3) Structure selection: Every 4 students form a group. They will choose one type of structure combining one or two materials from wood, steel and concrete for small and medium-span space training. The appropriate structure type can also be selected according to the needs of the project progress.

4) Selection of cladding materials: Pay attention to the coordination of roof materials and structure selection, as well as the insulation performance and anti-fogging of the outer protective material of the swimming hall.

5) Air conditioning method: The swimming hall and paddling pool adopt central air conditioning and mechanical ventilation. The rest of the rooms adopt VRV, and the use of functional rooms should consider the use of natural lighting and ventilation in spring and autumn and the transition seasons.

6) The use of Revit is encouraged to promote the curriculum of design.

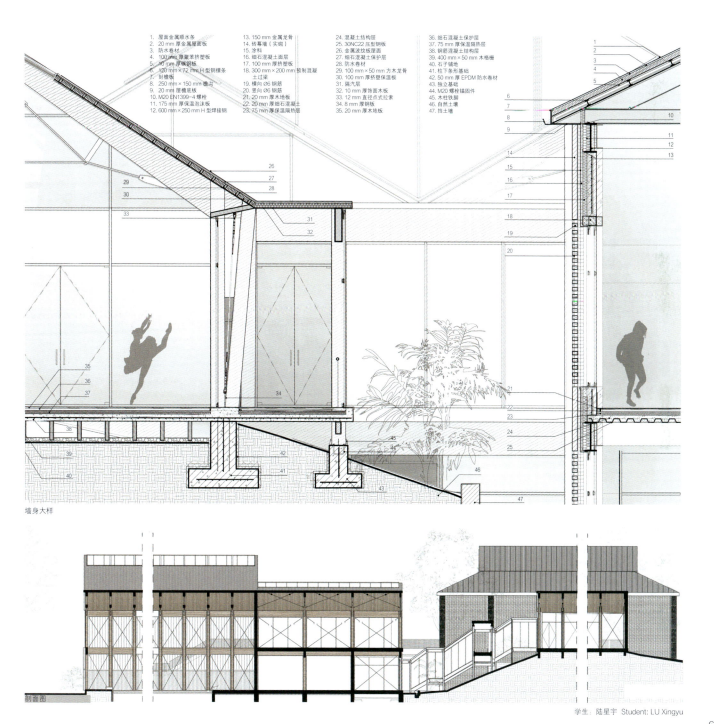

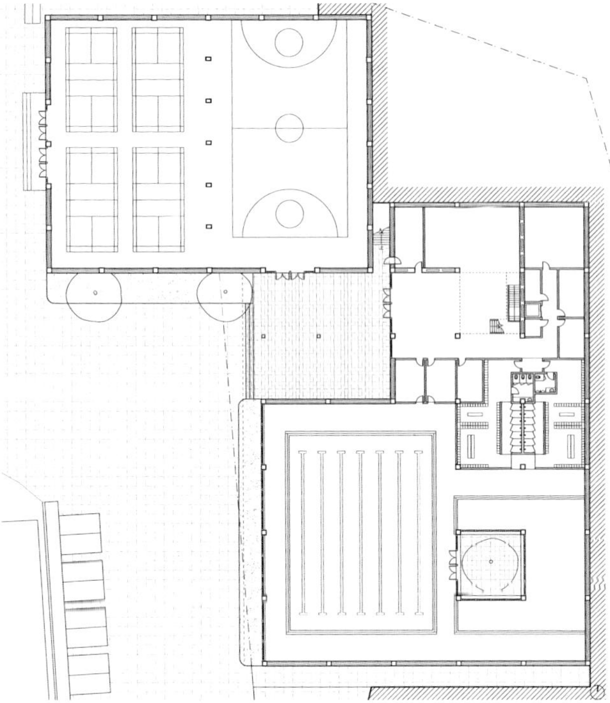

一层平面图 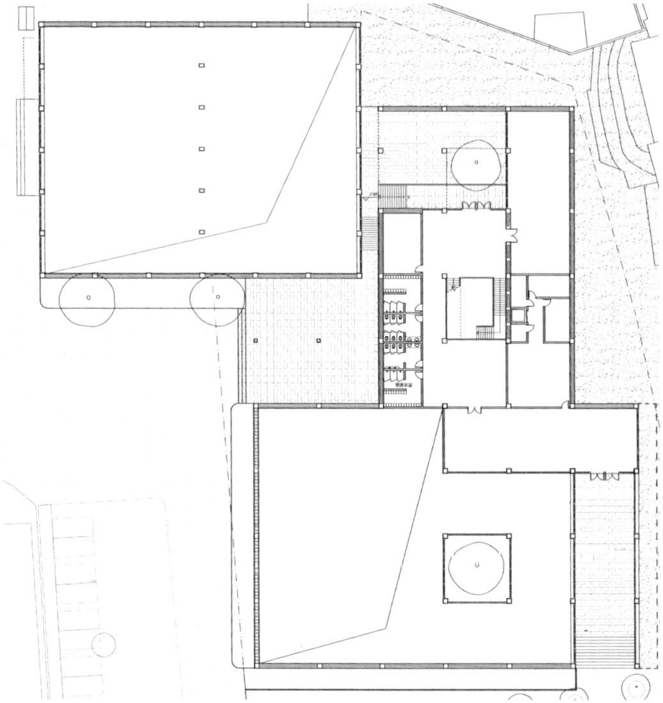　　　　　　　　　　　　　　　　　　　二层平面图

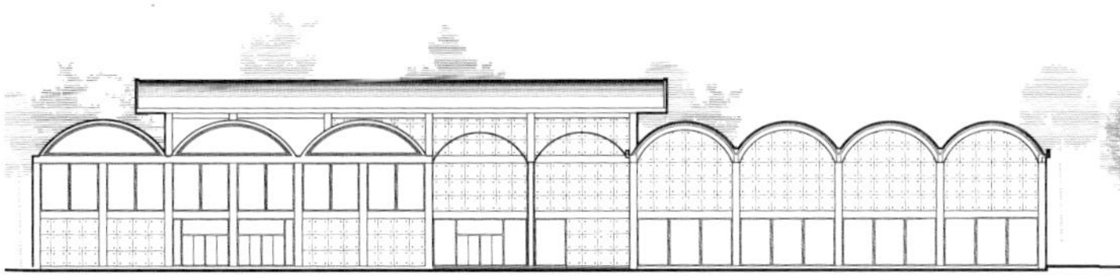

西立面图 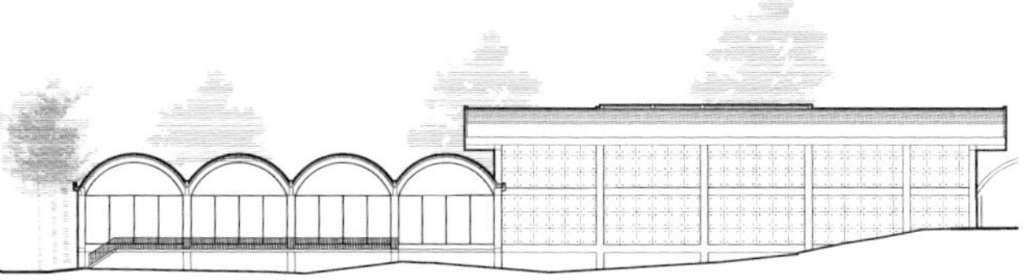　　　　　　　　　　　　　　　　　　　　东立面图

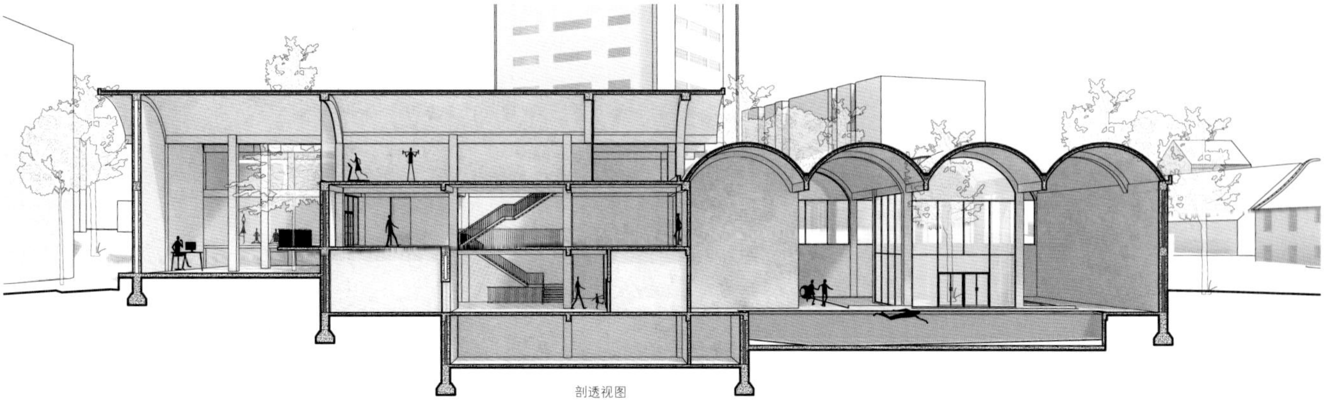

剖透视图

学生：沈至文　Student: SHEN Zhiwen

65

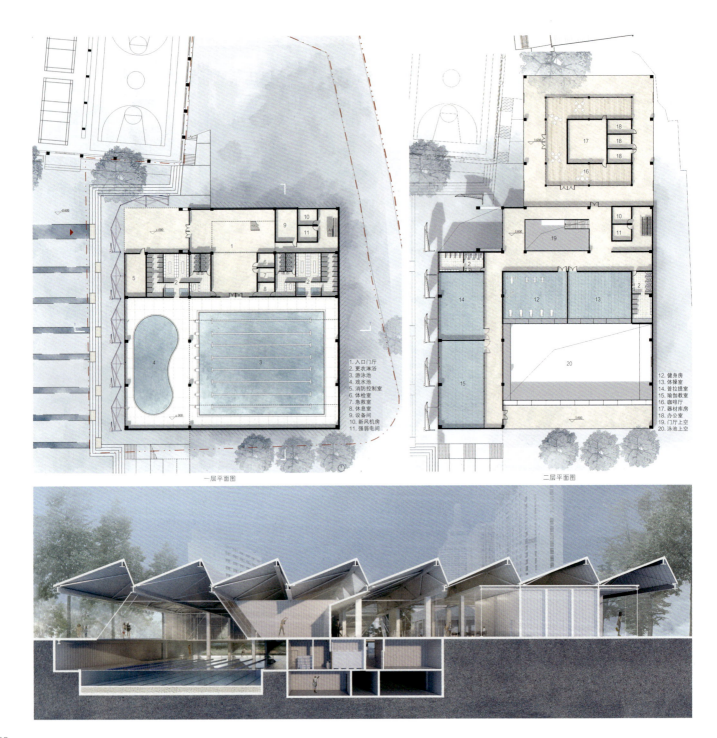

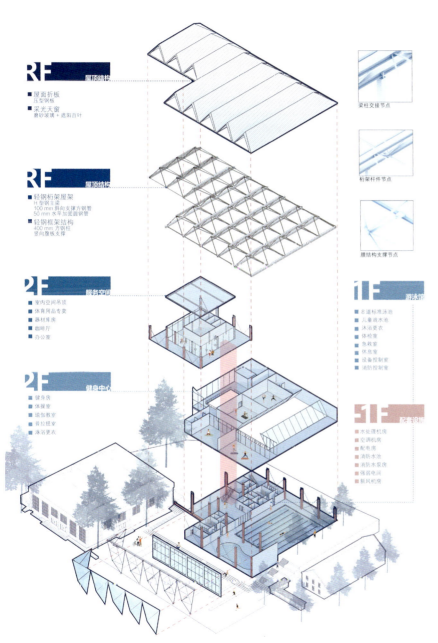

学生：李静怡 Student: LI Jingyi

社区文化艺术中心设计
DESIGN OF COMMUNITY CULTURE AND ART CENTER

张雷　王铠　尹航
ZHANG Lei　WANG Kai　YIN Hang

概况
本项目拟在百子亭历史风貌区基地处新建社区文化中心，总建筑面积约 8 000 m²，项目不仅为周边居民文化基础设施服务，同时也期望成为复兴老城的街区活力的文化地标。根据基地条件、功能使用进行建筑和场地设计。

民国时期，百子亭一带属于高级住宅区，在位置上紧邻作为文教区的鼓楼，以及作为市级行政区的傅厚岗地区。凭借区域上的优势与政府的扶持，百子亭一带自1930 年代开始，逐渐成为当时文化精英、社会名流与政府要员的聚集之地。众多受邀前往南京创建其事业的学者、文人都在此购买土地，并建造出了一幢幢"和而不同"的新式住宅。这些建筑既是近代南京城市肌理中的现代图景，也是当时中国有为之士们实现梦想的舞台，还是中国近现代建筑史中不可忽视的华美段落。

基地条件
根据《南京历史文化名城保护规划（2010—2020）》，百子亭历史风貌区被列入"保护名录"。百子亭历史风貌区内现有市级文物保护单位 3 处，为桂永清公馆旧址、徐悲鸿故居和傅抱石故居，不可移动文物 8 处，历史建筑 1 处。

设计内容
1）演艺中心。包含 400 座的小剧场，乙级。台口尺寸为 12 m×7 m。根据设计的等级确定前厅、休息厅、观众厅、舞台等面积。观众厅主要为小型话剧及戏剧表演而设置。按 60—80 人化妆布置化妆室及服装道具室，并设 2—4 间小化妆室。要求有合理的舞台及后台布置，应设有排练厅、休息室、候场厅以及道具存放间等设施，其余根据需要自定。

2）文化中心。定位于区级综合性文化站，包括公共图书阅览室、电子阅览室、多功能厅、排练厅以及辅导培训、书画创作等功能室（不少于 8 个且每个功能室面积应不低于 30 m²）。

3）配套商业。包含社区商业以及小型文创主题商业单元。其中社区商业为不小于 200 m² 超市一处，文创主题商业单元面积为 60—200 m²。

4）其他。变电间、配电间、空调机房、售票、办公、厕所等服务设施根据相关设计规范确定，各个功能区可单独设置，也可统一考虑。地上不考虑机动车停车配建，街区地下统一解决，但需要根据建筑功能面积计算数量。

教学成果
每人不少于 4 张 A1 图纸，图纸内容包括：
1）城市与环境：1∶500 总平面图，总体鸟瞰图、轴测图。
2）空间基本表达：1∶200—1∶400 平、立、剖面图。
3）空间解析与表现：概念分析图、空间构成分析图、轴测分析图、剖透视图（不少于 2 张，必须包含大空间、公共空间的剖透视图）、室内外人眼透视图若干。
4）手工模型：每个指导教师组内各做一个 1∶500 总图体量模型，每位学生做一个 1∶500 的概念体块模型。

教学进度
本次设计课程共 8 周。
第一周：授课（1 学时），调研场地及案例，制作场地模型（SU 模型 +1∶500 实体模型），收集相应的案例资料。
第二周：学生收集案例汇报、初步概念方案讨论（包含体块与场地关系布局、内部空间基本布局）。
第三周：概念深化，完善初步建筑功能布局和空间形态方案（包括基本空间单元及其组合），制作空间形态模型。
第四周：方案定稿，明确空间表皮、平面功能、街区环境模式。
第五周：方案深化 I，深化空间表皮、平面功能、街区环境。
第六周：方案深化 II，细化表皮处理、剧场空间及其他重要公共空间的设计。
第七周：方案表达，完成平、立、剖面图绘制，完善 SU 设计模型。
第八周：制图、排版。

Overview
The project plans to build a new community culture and art center at the base of Baiziting historical area, with a total construction area of about 8,000 m². The project not only serves the cultural infrastructure of surrounding residents, but also hopes to become a cultural landmark to revive the vitality of the old city. Design the building and site according to the base conditions and functions.

During the period of the Republic of China, Baiziting was a high-end residential area, close to the Gulou area as a cultural and educational area and Fuhougang area as a municipal administrative area. Relying on its regional advantages and government support, Baiziting area gradually became settlements for cultural elites, celebrities and government officials since the 1930s. Many scholars and literati invited to Nanjing to establish their careers bought land here and built new houses which were "harmonious but different". These buildings are not only the modern picture in the modern urban texture of Nanjing, but also the stage for

Chinese promising people to realize their dreams at that time, and the gorgeous paragraphs that cannot be ignored in the history of Chinese modern architecture.

Site conditions

According to Nanjing Historical and Cultural City Protection Plan (2010–2020), Baiziting Historical Landscape Protection Area is included in the "Protection List". There are 3 municipal cultural relics protection units in Baiziting Historical Landscape Protection Area, including the former site of GUI Yongqing residence, XU Beihong's former residence and FU Baoshi's former residence, 8 immovable cultural relics and 1 historical building.

Design content

1) Performance arts center. It contains 400-seat small theatre of Class B. The size of the proscenium is 12 m × 7 m. The area of front hall, lounge, auditorium and stage shall be determined according to the design level. The auditorium is mainly set up for small-scale drama and drama performances. The dressing room and costume room for 60–80 people and 2–4 small dressing rooms shall be arranged. Reasonable stage and backstage arrangements are required. Rehearsal hall, lounge, waiting area, props storage room and other facilities can be set, and the rest shall be determined according to needs.

2) Cultural center. It is positioned as a district-level comprehensive cultural station, including a public reading room, an electronic reading room, a multi-functional hall, a rehearsal hall, counseling and training classroom, calligraphy and painting space and other functional rooms (no less than 8 and the area of each functional room shall not be less than 30 m^2).

3) Supporting business. It includes community business and small cultural and creative theme business units. Among them, the community business is a supermarket with an area of no less than 200 m^2, and the area of cultural and creative theme business unit is 60–200 m^2.

4) Others. Service facilities such as substation rooms, power distribution rooms, air conditioning rooms, ticket offices, offices and toilets are determined according to relevant design specifications. Each functional area can be set separately or considered uniformly. The parking allocation of motor vehicles is not considered on the ground, and the underground of the block is solved uniformly, but the quantity needs to be calculated according to the building functional area.

Teaching achievements

Each person should produce at least 4 A1 drawings, including:

1) City and environment: 1∶500 general layout, overall aerial view, and axonometric drawing.

2) Basic spatial expression: 1∶200–1∶400 plans, elevations and sections.

3) Spatial analysis and expression: Concept analysis drawings, spatial composition analysis drawings, axonometric analysis drawings, sectional perspective views (no less than 2, which must include the sectional perspective view of large space and public space), indoor and outdoor human-eye perspective views.

4) Manual model: Each tutor group makes a 1∶500 general layout volume model, and each student makes a 1∶500 concept volume model.

Teaching schedule

This design course is 8 weeks in total.

Week 1: Teaching (1 class hour), researching sites and cases, making site models (SU model + 1∶500 physical model) and collecting relevant case information.

Week 2: Students report their collected cases and discuss preliminary conceptual plans (including the layout of the relationship between the block and the site, and the basic layout of the interior space).

Week 3: Concept deepening, completing the preliminary architectural functional layout and spatial form scheme (including basic spatial units and their combinations), and making a spatial form model.

Week 4: Finalizing the plan, and clarifying the space surface, plan function, and block environment pattern.

Week 5: Plan deepening I, deepening the space surface, plan function, and block environment.

Week 6: Plan deepening II, refining the design of the surface treatment, theater space and other important public spaces.

Week 7: Schematic expression, completing the plan, elevation and section, and improving SU design model.

Week 8: Drawing and layout.

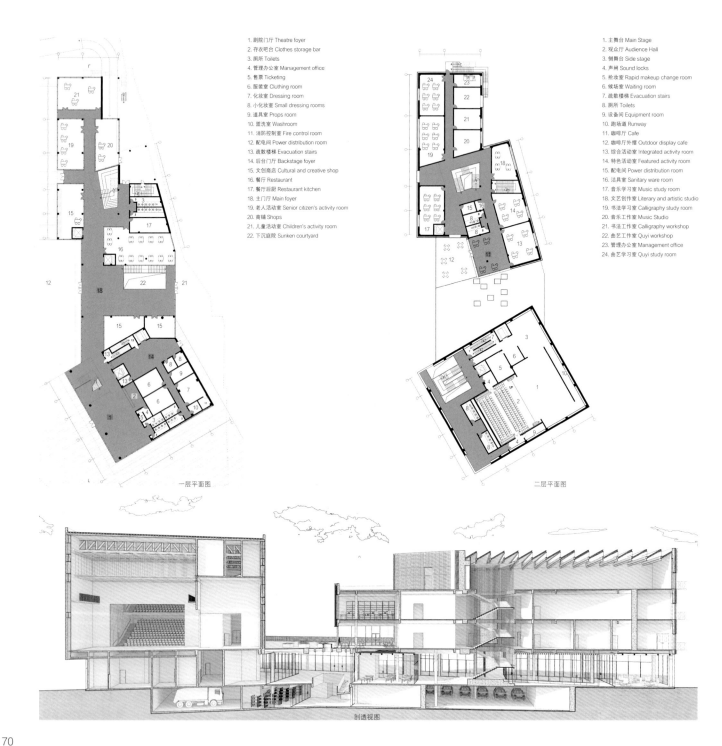

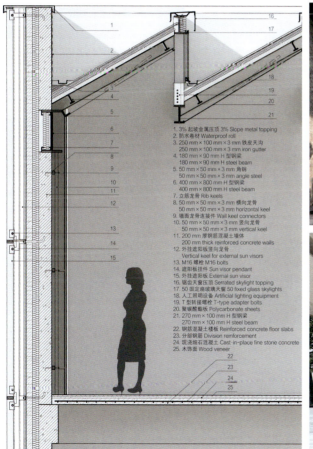

1. 3% 起坡金属压顶 3% Slope metal topping
2. 防水卷材 Waterproof roll
3. 250 mm × 100 mm × 3 mm 铁皮天沟
   250 mm × 100 mm × 3 mm iron gutter
4. 180 mm × 90 mm H 型钢梁
   180 mm × 90 mm H steel beam
5. 50 mm × 50 mm × 3 mm 角钢
   50 mm × 50 mm × 3 mm angle steel
6. 400 mm × 800 mm H 型钢梁
   400 mm × 800 mm H steel beam
7. 立筋龙骨 Rib keels
8. 50 mm × 50 mm × 3 mm 横向龙骨
   50 mm × 50 mm × 3 mm horizontal keel
9. 墙面龙骨连接件 Wall keel connectors
10. 50 mm × 50 mm × 3 mm 竖向龙骨
    50 mm × 50 mm × 3 mm vertical keel
11. 200 mm 厚钢筋混凝土墙体
    200 mm thick reinforced concrete walls
12. 外挂钢板竖向龙骨
    Vertical keel for external sun visors
13. M16 螺栓 M16 bolts
14. 遮阳板挂件 Sun visor pendant
15. 外挂遮阳板 External sun visor
16. 锯齿天窗压顶 Serrated skylight topping
17. 50 固定玻璃天窗 50 fixed glass skylights
18. 人工照明设备 Artificial lighting equipment
19. T 型转接螺栓 T-type adapter bolts
20. 聚碳酸酯板 Polycarbonate sheets
21. 270 mm × 100 mm H 型钢梁
    270 mm × 100 mm H steel beam
22. 钢筋混凝土楼板 Reinforced concrete floor slabs
23. 分部钢筋 Division reinforcement
24. 现浇细石混凝土 Cast-in-place fine stone concrete
25. 木饰面 Wood veneer

学生：陆星宇，刘卓然　Students: LU Xingyu, LIU Zhuoran

2-2 剖立面图　　　　　　南立面图

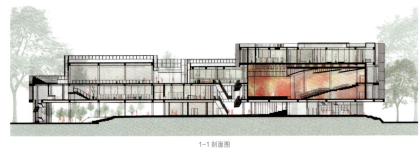

1-1 剖面图

西立面图

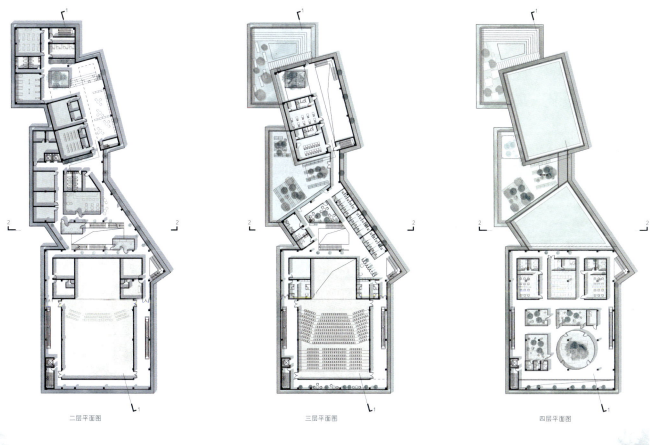

二层平面图　　　　　三层平面图　　　　　四层平面图

学生：李静怡，李沛熹　Students: LI Jingyi, LI Peixi

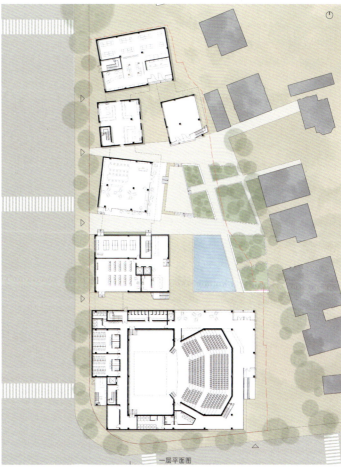

一层平面图

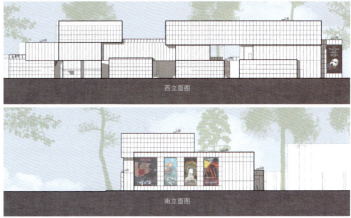

西立面图

南立面图

学生：张伊儿，杨曦睿　Students: ZHANG Yi'er, YANG Xirui

建构设计研究 CONSTRUCTION DESIGN RESEARCH

工 地 实 习
CONSTRUCTION SITE INTERNSHIP
傅筱　吴佳维
FU Xiao　WU Jiawei

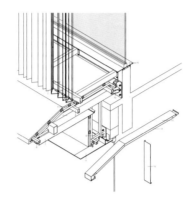

教学目标
　　工地实习的训练目的是加深学生对建筑从图纸到实际建造过程的认识和理解，重点理解图纸与建造的关联。本课程的任务主要是通过工地现场考察，让学生了解建筑材料的生产过程，对建筑实际建造流程有一定的直观认识；让学生实地观察某一建筑构件的建造过程，绘制建筑构件施工图纸；让学生掌握建筑面层材料与建筑主体结合的构造关系，理解建筑材料运用与设计表达之间的关联性。

人员组成
　　学生：2020级24人，实际参加人数24人。
　　任课教师：傅筱教授；吴佳维副研究员。
　　现场指导：建材厂技术人员。
　　助教：研究生陈雯。

实践时间及地点
　　实习时间：2023年6月至9月，建材厂参观2天，研究及绘图约10天。
　　参观地点：科信杰铝业科技（常州）有限公司、上海凡柏建筑科技有限公司、涂耐可艾克（上海）涂料有限公司。

教学内容
　　课堂讲授：由教师课堂讲授相关的技术知识。
　　厂家调研：教师和助教带领学生在建材厂进行考察和学习。
　　知识梳理：结合调研收集到的资料和教师的讲解，梳理从原材料到建材生产、现场安装的全过程。
　　构造轴测图绘制：进一步搜索图集、案例，绘制相关节点的构造轴测图。

　　实习报告：根据厂家考察和课堂讲授的学习，完成实习报告。

教学成果
　　工地实习是南京大学建筑与城市规划学院建筑学本科教学中继"建筑技术（一）建构设计"课程与"古建筑测绘"课程后对建造议题的又一次专题训练。自2022年以来，课程的调研对象从工地拓展至一些具备先进研发和生产技术的建筑材料企业。学生首先从生产环节现场考察建材的制作工艺和特性，向现场技术人员了解施工基本流程；其次，进一步查找资料，完成"原材料—产品—生产流程—现场施工顺序和标准节点"的知识梳理，深入学习建筑材料在建成建筑中的运用；最后，在有限的图纸和照片资料基础上，通过拓展资料学习、节点原理推导、课堂教学研讨三个环节，以BIM建模的方式再现建筑表皮与承重结构相遇的关键技艺。

Teaching objectives
　　The training purpose of on the construction site internship is to deepen students' knowledge and understanding of the process from architectural drawings to actual construction with a focus on understanding the connection between drawings and construction. The main tasks of this course are as follows: Let students understand the production process of building materials and have a certain intuitive understanding of the actual construction process of buildings through on-site inspection; let students observe the construction process of a specific building component on the spot and draw construction drawings of the components; let students master the structural relationship between surface

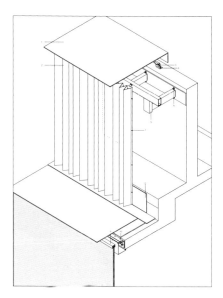 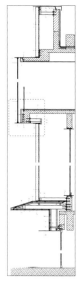 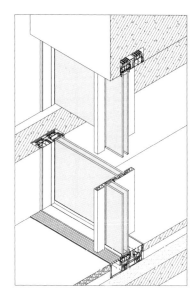 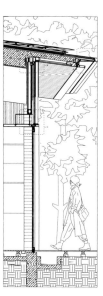

materials and the main body of the building, and understand the correlation between the use of building materials and design expression.

Staff composition

Students: 24 students from the class of 2020, the actual number of participants is 24.
Teachers: Professor Fu Xiao; Associate Researcher Wu Jiawei.
On-site guidance: technical staff of the building material factory.
Teaching assistant: Chen Wen, graduate student.

Internship time and place

Internship time: From June to September 2023, visit to the building material factory lasting for 2 days, and research and drawing for about 10 days.
Visiting places: Kexinjie Aluminum Technology (Changzhou) Co., Ltd., Shanghai VBM Architectural Technology Co., Ltd., and Terraco Coating (Shanghai) Co., Ltd..

Teaching content

Classroom lectures: Teachers will give lectures on relevant technical knowledge in class.
Factory research: Teachers and the teaching assistant will lead students to conduct inspections and study in building material factories.
Knowledge combing: Comb the whole process from raw materials to building material production and on-site installation in combination with the information collected in the research and the teacher's explanation.
Structural axonometric drawing: Further search for atlases and cases, and draw structural axonometric drawings of relevant nodes.
Internship report: According to the manufacturer's inspection and classroom lectures, complete the internship report.

Teaching achievements

This construction site internship is another special training on construction issues in the undergraduate teaching of architecture in the School of Architecture and Urban Planning of Nanjing University following the courses of "Architectural Technology 1: Construction Design" and "Surveying and Mapping of Ancient Buildings". Since 2022, the research objects of the course have been expanded from the construction site to some building material companies with advanced research and development and production technologies. Students will first inspect the production process and the characteristics of building materials from the production chain, and understand the basic construction process from on-site the technical staffs. Then students will search for further information and complete the knowledge combination of "raw materials—products—production process—on-site construction sequence and standard nodes", and learn the application of materials in use of building construction. Based on limited drawings and photos, through the expansion of information, principal derivation and classroom teaching and discussion, finally the reproduce of the key technology between building surface and load-bearing structure are shown in the way of BIM modeling.

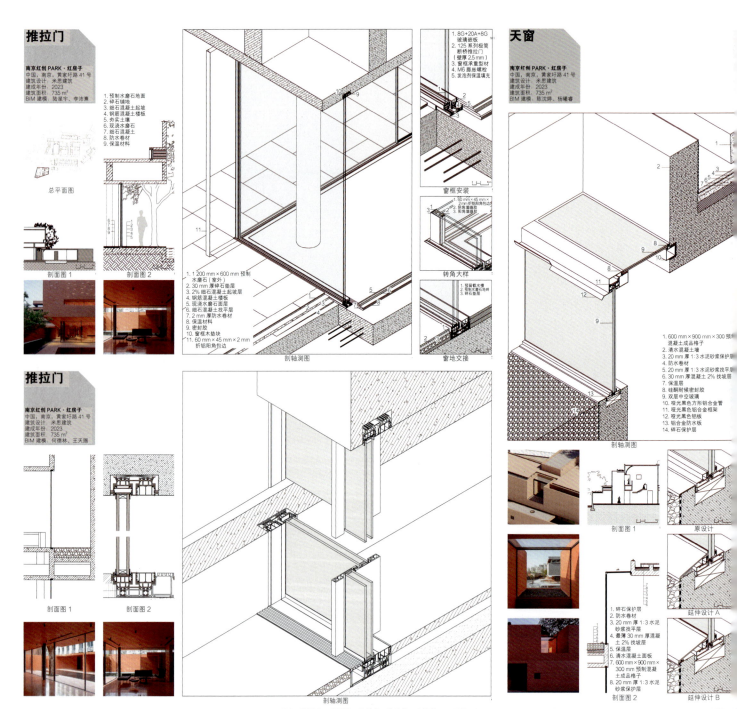

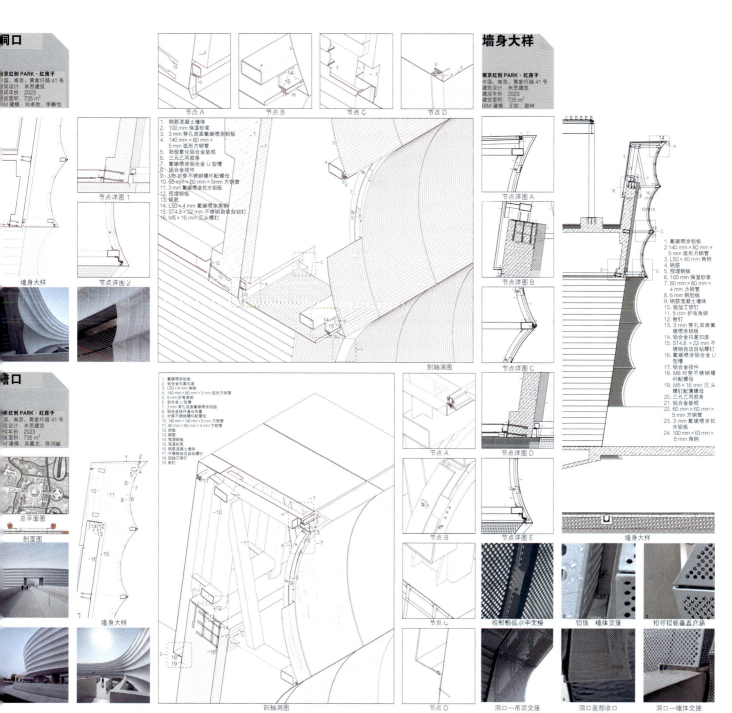

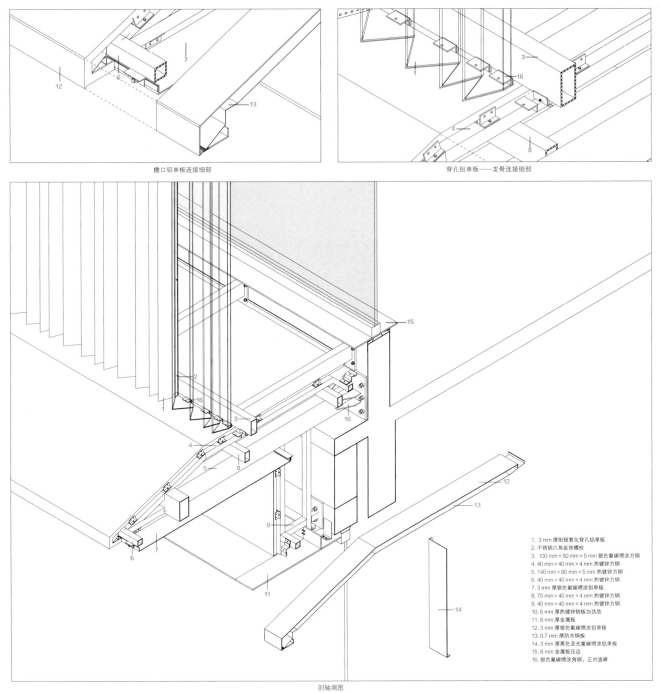

檐口铝单板连接细部

穿孔铝单板——龙骨连接细部

1. 3 mm 厚阳极氧化穿孔铝单板
2. 不锈钢六角装饰螺栓
3. 100 mm × 50 mm × 5 mm 银色氟碳喷涂方钢
4. 40 mm × 40 mm × 4 mm 热镀锌方钢
5. 140 mm × 80 mm × 5 mm 热镀锌方钢
6. 40 mm × 40 mm × 4 mm 热镀锌方钢
7. 3 mm 厚银色氟碳喷涂铝单板
8. 70 mm × 40 mm × 4 mm 热镀锌方钢
9. 40 mm × 40 mm × 4 mm 热镀锌方钢
10. 6 mm 厚热镀锌钢板加劲肋
11. 8 mm 厚金属板
12. 3 mm 厚银色氟碳喷涂铝单板
13. 0.7 mm 厚防水钢板
14. 3 mm 厚黑色亚光氟碳喷涂铝单板
15. 8 mm 金属板压边
16. 银色氟碳喷涂角钢，正对波峰

剖轴测图

学生：李若松，陈玙 Students: LI Ruosong, CHEN Cheng

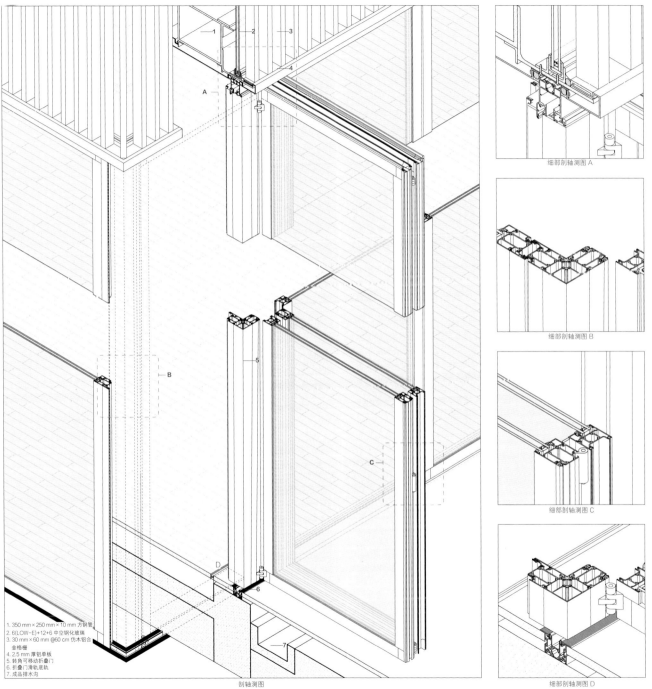

1. 350 mm × 250 mm × 10 mm 方钢管
2. 6(LOW-E)+12+6 中空钢化玻璃
3. 30 mm × 60 mm @60 cm 仿木铝合金格栅
4. 2.5 mm 厚铝单板
5. 转角可移动折叠门
6. 折叠门滑轨底轨
7. 成品排水沟

剖轴测图

细部剖轴测图 A

细部剖轴测图 B

细部剖轴测图 C

细部剖轴测图 D

学生：沈至文，高晴　Students: SHEN Zhiwen, GAO Qing

建筑认知实践 ARCHITECTURAL COGNITIVE PRACTICE

古 建 筑 测 绘
SURVEYING AND MAPPING OF ANCIENT BUILDINGS

赵辰 史文娟 赵潇欣
ZHAO Chen SHI Wenjuan ZHAO Xiaoxin

育人目标
课程践行习近平总书记关于传承历史文化、保护城市文化遗产的重要讲话精神，着重进行对历史珍贵遗产的抢救性测绘，令同学们感受到社会责任感和对传统文化保护的使命感，有力引导学生树立文化自信、文化自觉、文化自强的新时代建筑理念；让学生了解设计及建造过程，浅尝匠人之心，在亲身劳动中体会知行合一；培养学生理解、认同并掌握中国传统建筑规划与建筑设计的核心理念，充分认识建筑遗产价值，为建筑学后续学习打下坚实基础。

教学目标
本课程是建筑学专业本科生的专业基础理论课程。其任务是使学生切实理解中国传统建筑结构体系、构造关系及比例尺度等基本概念，培养学生对传统建筑年代鉴定和价值判断的基本技能。

与学校人才培养的契合关系
培养学生广博的知识技能以及卓越的专业素养，培养学生合作精神与领导能力，培育学生强烈的家国情怀与社会责任感，培养学生有效沟通的能力。

教学内容
通过室外作业和室内工作两个阶段，完成现场测绘和整理图纸报告两个环节；"测"，观测量取现场实物的尺寸数据；"绘"，根据测量数据与草图整理绘制完备的测绘图纸，最终完成个人独立的答辩报告。

第一阶段：抵达现场，听讲座（第1天）。
第二阶段：现场工作，各小组分工协作（第2—7天）。
第三阶段：图纸绘制，小组资料归档、排版（第8—14天）。
第四阶段：答辩准备，打印图纸、个人独立研究（第15—20天）。
第五阶段：交图答辩（第21天）。

Educational goals
The course implements the spirit of General Secretary Xi Jinping's important speech on inheriting history and culture and protecting urban cultural heritage, focusing on the rescue survey and mapping of precious historical heritage, making students feel a sense of social responsibility and a sense of mission for the protection of traditional culture, and effectively guiding students establish a new era of architectural concepts of cultural self-confidence, cultural awareness, and cultural improvement; let students understand the design and construction process, experience the craftsman's heart, and realize the unity of knowledge

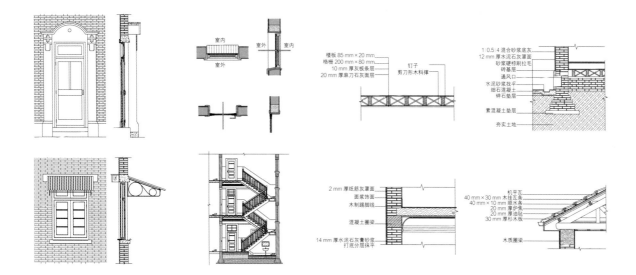

and action through personal labor; cultivate students' understanding, recognition, and mastery of the core concepts of traditional Chinese architectural planning and architectural design, and fully understand the value of architectural heritage and lay a solid foundation for subsequent study of architecture.

Teaching objectives

This course is a professional basic theory course for undergraduates majoring in architecture. The mission of this course is to enable students to truly understand the basic concepts such as the building structural system, structural relationship, and proportional scale of traditional Chinese buildings, and to cultivate students' basic skills in age identification and value judgment of traditional buildings.

Compatible relationship with school talent training

Cultivate students' extensive knowledge and skills, develop students' excellent professional qualities, foster students' cooperative spirit and leadership skills, cultivate students' strong feelings for family and country and social responsibility, and cultivate students' ability to communicate effectively.

Teaching content

Through the two stages of outdoor work and indoor work, the two links of on-site surveying and mapping and drawing report preparation are completed: "Measuring", observing and measuring the dimensional data of the actual objects on site; "Drawing", organizing and drawing complete surveying and mapping drawings based on the measurement data and sketches, and finally completing a personal and independent defense report.

Phase 1: Arriving at site and attending lectures (Day 1).

Phase 2: On-site work, division of labor and collaboration among each group (Days 2 to 7).

Phase 3: Drawing drawings, archiving and typesetting of group materials (Days 8 to 14).

Phase 4: Defense preparation, printing drawings and independent research (Days 15—20).

Phase 5: Submission defense (Day 21).

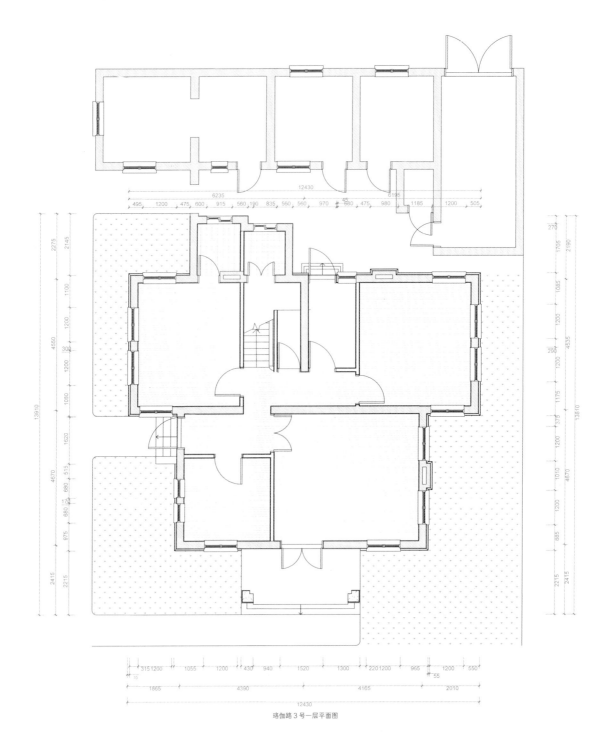

珞伽路3号一层平面图

珞伽路3号北立面图

学生：李沛熹，黄淑睿，陈沈婷，吴嘉文　Students: LI Peixi, HUANG Shurui, CHEN Shenting, WU Jiawen

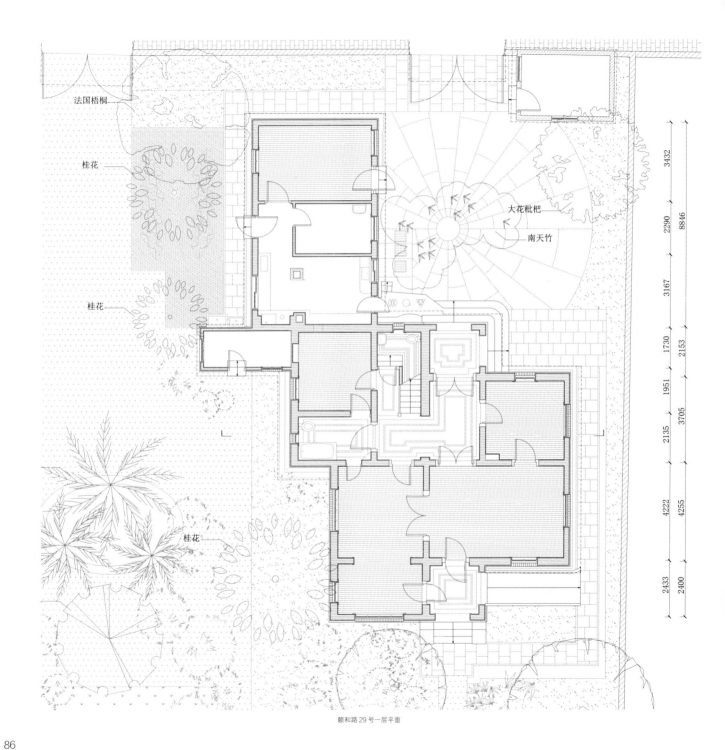

颐和路29号一层平面

颐和路29号剖面图

学生：李静怡，杨曦睿，张伊儿，陆星宇 Students: LI Jingyi, YANG Xirui, ZHANG Yi'er, LU Xingyu

建筑设计（七）ARCHITECTURAL DESIGN 7

高层办公楼设计
DESIGN OF HIGH-RISE OFFICE BUILDINGS

尹航　王铠
YIN Hang　WANG Kai

教学目标

建筑设计七共分为两组，学生们可以选择不同的建筑设计概念发展方向：性能化导向建筑设计、景观导向的高层建筑、复合功能的立体城市等。

本组设计选题既需要研究复合的新型建筑功能，也需要对周边城市生态和景观进行合理回应，力求在保留高层建筑的复杂性基础上，增加城市景观、建筑生态化的要求。

本次课程设计首先希望学生了解当代建筑设计行业中高层建筑的基本特点，研究当代高层建筑的设计策略，了解高层建筑涉及的相关规范与知识，提高综合分析及解决问题的能力。其次还希望学生能主动将建筑与生态、景观等环境要素有机结合，研究环境，在建筑方案中预设某种策略，创造出从概念、策略到解决方案一体的高层建筑设计。

教学内容

1）场地与相关指标

本次课程的场地位于南京市高新区南京软件园的中部核心区位置。地块面积约 1.7 hm^2，要求设计一处集生产办公、人才公寓、商业配套、设备用房等于一体的复合功能建筑。建筑容积率控制为 3.0，限高 100 m，绿地率 30%。其他具体指标可参考《〈南京市城市规划条例〉实施细则》等。

2）功能与交通

设计地块在上位规划中确定的性质为科研设计用地，这是近两年新设定的城市建设用地性质，常用于科技园区等新城发展用地。其具体功能相对灵活，可承载生产、办公、公寓、服务等相关复合功能。学生们须调研相关案例，并自主策划本地块的具体功能。具体面积比例在合理的基础上灵活自定。地块需考虑周边地块与交通等情况，合理组织车行、人行流线，按不同功能设置出入口集散空间；合理设置场地交通与周边道路关系，合理设置停车场。停车数量（机动车与非机动车）应尽量满足地方法规要求，地下停车场出入口大于2个。

3）相关规范

《民用建筑设计统一标准》（GB 50352—2019）

《办公建筑设计标准》（JGJ/T 67—2019）

《车库建筑设计规范》（JGJ 100—2015）

《建筑设计防火规范（2018 年版）》（GB 50016—2014）

《汽车库、修车库、停车场设计防火规范》（GB 50067—2014）

《南京市建筑物配建停车设施设置标准与准则（2019 年版）》

时间安排（8 周）

第一周：场地分析（1 : 1 000 场地环境模型）及设计策略研究。

第二周：设计策略推敲（1 : 1 000 模型）。

第三周：概念模型制作（1 : 1 000 模型）。

第四周：总平面设计（草图、1 : 500 草模）。

第五周：平面图、立面图与细部深化设计。

第六周：技术图纸制作、排版。

第七周：技术图纸完成，表现图制作，表现模型制作。

第八周：表现图完成，模型完成。

Teaching objectives

Architectural Design 7 is divided into two groups. Students can choose different development directions of architectural design concepts: performance-oriented architectural design, landscape-oriented high-rise buildings, multi-functional three-dimensional cities, etc.

The design topic of this group requires both the study of new composite building functions and a reasonable response to the surrounding urban ecology and landscape. On the basis of retaining the complexity of high-rise buildings, the requirements for urban landscape and architectural ecology are added.

The design of this course firstly hopes that students will understand the

basic characteristics of high-rise buildings in the contemporary architectural design industry, study the design strategies of contemporary high-rise buildings, understand the relevant specifications and knowledge involved in high-rise buildings, and improve their ability to comprehensively analyze and solve problems. Secondly, we also hope that students can take the initiative to organically combine architecture with environmental elements such as ecology and landscape, study the environment, and preset certain design strategies in the architectural plan to create a high-rise building design that integrates concepts, strategies, and solutions.

Teaching content

1) Site and related indicators

The site for this course is located in the central core area of Nanjing Software Park in the High-tech Zone of Nanjing City. The plot area is about 1.7 hm^2, and it is required to design a composite functional building integrating production offices, talent apartments, commercial facilities, equipment rooms and so on. The building floor area ratio is controlled to 3.0, the height limit is 100 m, and the green space rate is 30%. For other specific indicators, please refer to the *Implementation Rules of Nanjing Urban Planning Regulations* and so on.

2) Function and transportation

The nature of the designed plot determined in the upper-level planning is scientific research and design land. This is a newly set nature of urban construction land in the past two years. It is often used for development of new cities such as science and technology parks. Its specific functions are relatively flexible and can host production, offices, apartments, services and other related composite functions. Students need to research relevant cases and independently plan the specific functions of the local area. The specific area ratio can be flexibly determined on a reasonable basis. The plot needs to consider surrounding plots and traffic conditions, rationally organize vehicle and pedestrian circulation lines, set up entrance and exit distribution spaces according to different functions; rationally organize the relationship between site traffic and surrounding roads, and rationally set up parking lots. The number of parking lots (motor vehicles and non-mobile vehicles) should try to meet the requirements of local regulations, and there should be more than 2 entrances and exits to the underground parking lot.

3) Relevant specifications

Uniform Standard for Design of Civil Building (GB 50352—2019)
Standard for Design of Office Building (JGJ/T 67—2019)
Code for the Design of Parking Garage Buildings (JGJ 100—2015)
Code for Fire Protection Design of Buildings (2018) (GB 50016—2014)
Code for Fire Protection Design of Garage, Motor Repair Shop and Parking Area (GB 50067—2014)
Standards and Guidelines for the Installation of Parking Facilities in Buildings of Nanjing Municipality (2019)

Schedule (8 weeks)

Week 1: Site analysis (1∶1,000 site environment model) and design strategy research.
Week 2: Design strategy review (1∶1,000 model).
Week 3: Conceptual model construction (1∶1,000 model).
Week 4: General plan design (sketch, 1∶500 draft sketch).
Week 5: Deepening design of plans, elevations and details.
Week 6: Technical drawing production and layout.
Week 7: Technical drawing completed, presentation drawing made, and presentation model made.
Week 8: Presentation map completed, and model completed.

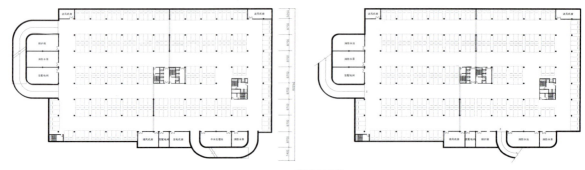

地下车库平面图

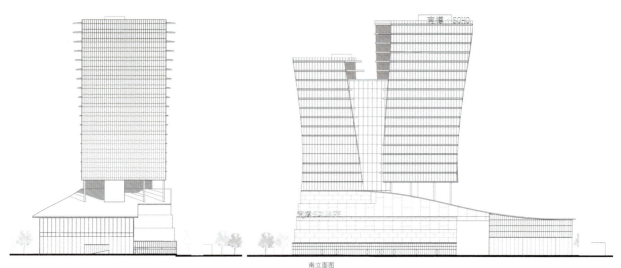

南立面图

1. Ø=20 mm 钢拉索
2. 5 mm×90 mm 遮阳金属百叶，间距 90 mm
3. 三层保温隔热玻璃窗
4. 不锈钢框架构件
5. 保温材料
6. 铰接金属连接件
7. 金属预埋件
8. 50 mm 厚硬木地板
9. 80 mm 厚轻质发泡混凝土保温材料
10. 80 mm 厚置热水管砂浆层
11. 490 mm 厚钢筋混凝土厚板结构层

1. Ø=20 mm steel cable
2. 5 mm×90 mm sunshade metal louvers, spacing 90 mm
3. Three-layer thermal insulation glass window
4. Stainless steel frame members
5. Insulation materials
6. Hinged metal connectors
7. Metal embedded parts
8. 50 mm thick hardwood floor
9. 80 mm thick lightweight foamed concrete insulation material
10. 80 mm thick built-in hot water pipe mortar layer
11. 490 mm thick reinforced concrete thick-slab structure layer

细部节点构造

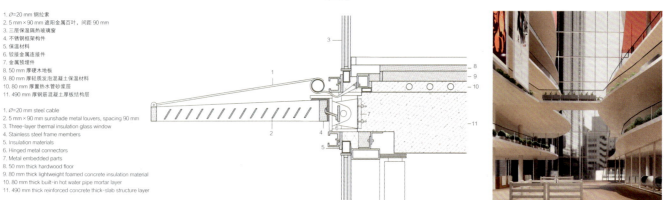

学生：邱雨婷，黄辰逸　Students: QIU Yuting, HUANG Chenyi

一层平面图

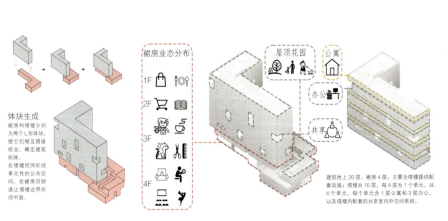

体块生成

裙房和塔楼分别为两个L形体块，使它们相互搭接咬合，确定建筑形体。
在塔楼挖洞形成单元性的公共空间，在裙房顶部退让塔楼边界形成中庭。

建筑地上20层，裙房4层，主要为塔楼提供配套设施；塔楼共16层，每4层为1个单元，共4个单元，每个单元含1层公寓和3层办公，以及塔楼内配套的共享室内外空间系统。

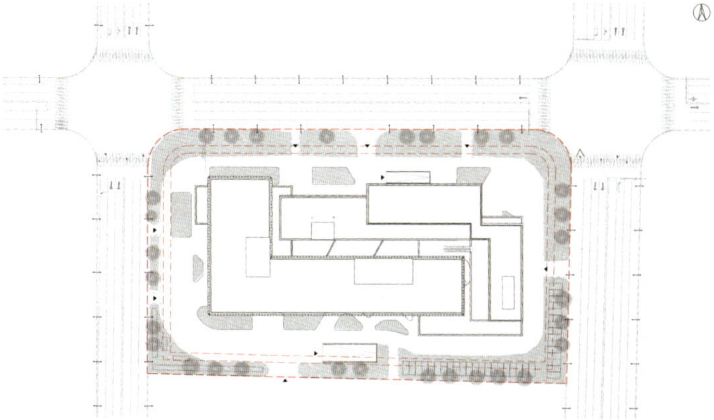

总平面图

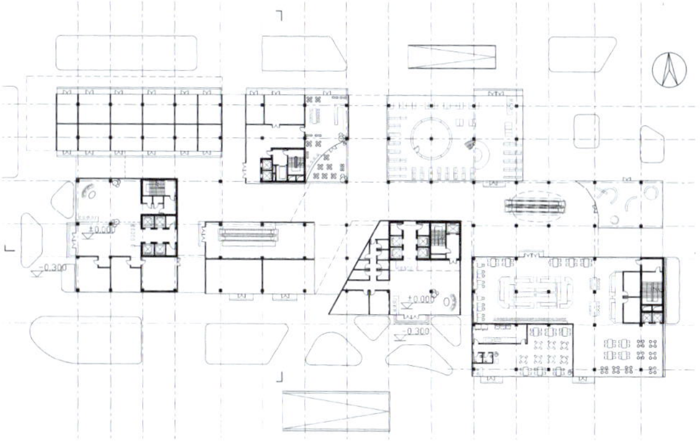

一层平面图

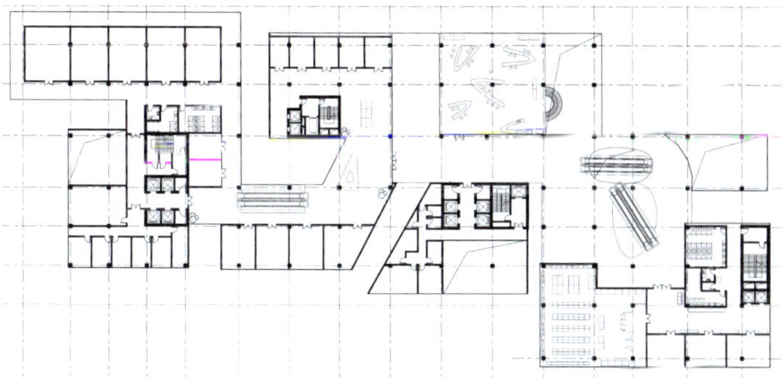

二层平面图

办公层平面图

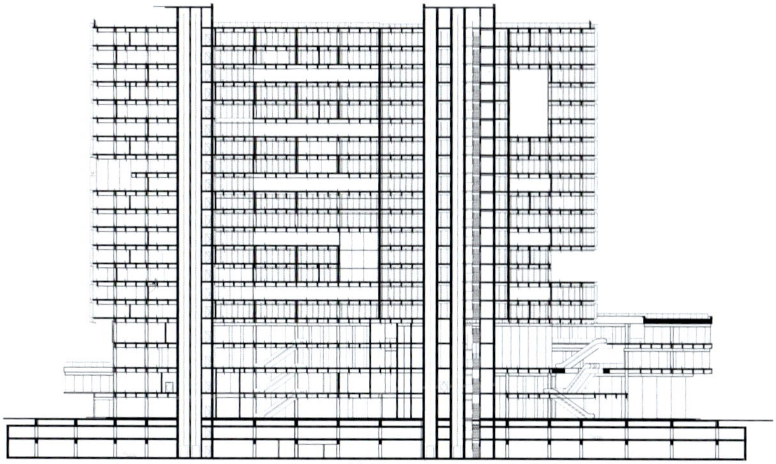

剖面图

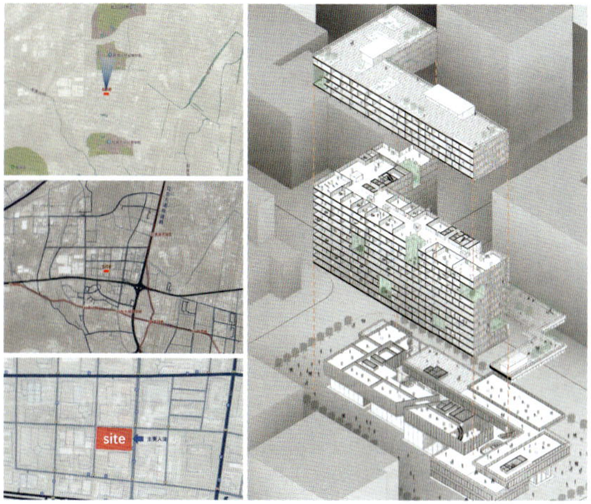

学生：高禾雨，唐诗诗 Students: GAO Heyu, TANG Shishi

93

城市设计
URBAN DESIGN

童滋雨 唐莲
TONG Ziyu TANG Lian

教学目标

中国的城市发展已经逐渐从增量扩张转向存量更新。通过对城市建成环境的更新改造提升环境性能和质量,将成为城市建设的新热点和新常态。与此同时,5G、物联网、无人驾驶等技术的发展又给城市环境的使用方式带来了新的变化。如何在城市更新设计中拓展建筑设计的边界也就成为新的挑战。

城市更新不但需要对建成环境本身有更充分的认知,也要对其中的人流、车流乃至水流、气流等各种动态的活动有正确的认知。从设计上来说,这也大大提高了设计者所面临的问题的复杂性,仅靠个人的直观感受和形式操作难以保证设计的合理性。而借助空间分析、数据统计、算法设计等数字技术,我们可以更好地认知城市形态的特征,理解城市运行的规则,并预测城市未来的发展。通过规则和算法来计算生成城市也是对城市设计思维范式的重要突破。

因此,本次设计将针对这些发展趋势,以城市街巷空间为研究对象,通过思考和推演探索其更新改造的可能性。通过本次设计,学生们可以理解城市设计的相关理论和方法,掌握分析城市形态和创造更好城市环境质量的方法。

设计场地

项目用地位于南京市鼓楼区湖南路街道青岛路社区陶谷新村,东至汉口路小区,西至上海路与汉口西路交叉口,属于南京市具有历史文脉的建成社区。项目总用地面积约 4 hm^2,具体设计范围可根据需要适当调整。

成果要求

本次设计以小组为单位,每小组 2 人。每组成果包括 8 张 A1 图纸和 1 份 A4 成果文本,具体内容可包括但不限于以下部分:

1)设计表达:平面图、立面图、轴测图、透视图等;
2)设计推演:设计形成过程的分析图;
3)设计评估:对设计成果的各种评估分析。

教学进度

阶段一: 场地调研与案例分析(2 周)。
阶段二: 设计目标确定与总体布局方案(2 周)。
阶段三: 方案完善与局部深化(3 周)。
阶段四: 制图与排版(1 周)。

Teaching objectives

China's urban development has gradually shifted from incremental expansion to stock renewal. Improving environmental performance and quality through the renewal and transformation of urban built environment will become a new hotspot and a new normal of urban construction. At the same time, the development of 5G, Internet of Things, unmanned driving and other technologies has brought new changes to the use of urban environment. How to expand the boundary of architectural design in urban renewal design has become a new challenge.

Urban renewal needs not only a better understanding of the built environment itself, but also a correct understanding of the pedestrian and vehicle flows, water and air currents and other dynamic activities. In terms of design, it also greatly

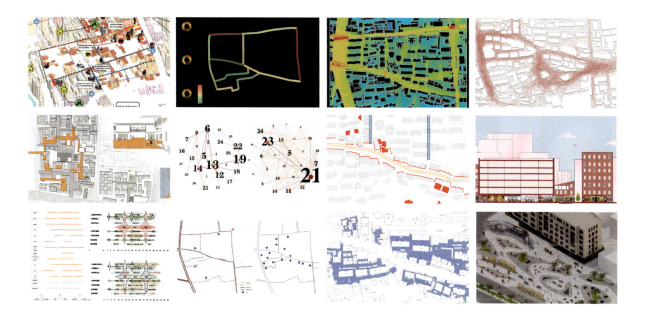

improves the complexity of the problems faced by designers. It is difficult to ensure the rationality of design only by personal intuitive feelings and formal operations. With the help of various digital technologies such as spatial analysis, data statistics and algorithm design, we can better understand the characteristics of the urban form, understand the rules of the urban operation, and predict the future development of the city. Calculating and generating cities through rules and algorithms is also an important breakthrough in the thinking paradigm of urban design.

Therefore, this design will aim at these development trends, take the urban street space as the research object, and explore the possibility of its renewal and transformation through thinking and deduction. Through this design, students can understand the relevant theories and methods of urban design, and master the methods of analyzing the urban form and creating better urban environmental quality.

Design site

The project site is located in Taogu New Village, Qingdao Road Community, Hu'nan Road Street, Gulou District, Nanjing City. It stretches from Hankou Road Community in the east to the intersection of Shanghai Road and Hankou West Road in the west, and belongs to a built community with a historical context in Nanjing City. The total land area of the project is about 4 hm^2, and the specific design scope can be adjusted as needed.

Achievement requirements

This design is carried out in groups, with 2 people in each group. The achievements each group include 8 A1 drawings and 1 A4 achievement text. The specific contents may include, but are not limited to, the following parts:

1) Design expression: Plans, elevations, axonometric drawings, perspective views, etc.

2) Design deduction: Analysis diagrams of design formation process.

3) Design evaluation: Various evaluation and analysis of design results.

Teaching progress

Stage 1: Site investigation and case analysis (2 weeks).

Stage 2: Determination of design objectives and overall layout scheme (2 weeks).

Stage 3: Scheme improvement and partial deepening (3 weeks).

Stage 4: Drawing and layout (1 week).

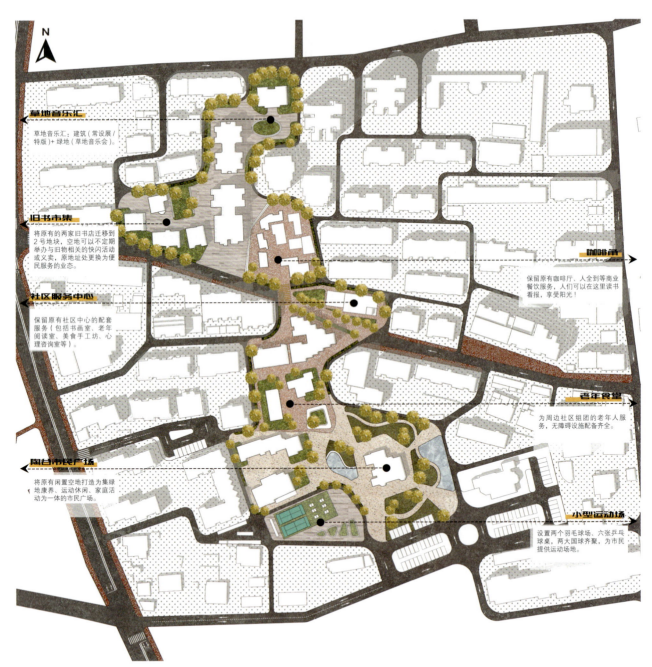

学生：邱雨婷，唐诗诗 Students: QIU Yuting, TANG Shishi

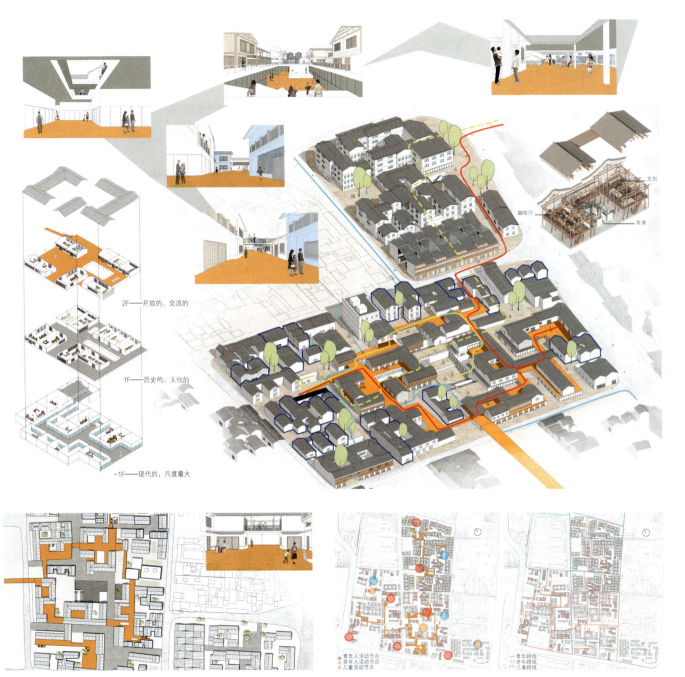

学生：黄辰逸，高禾雨，麦吾兰江·穆合塔尔　Students: HUANG Chenyi, GAO Heyu, Maiwulanjiang MUHETAR

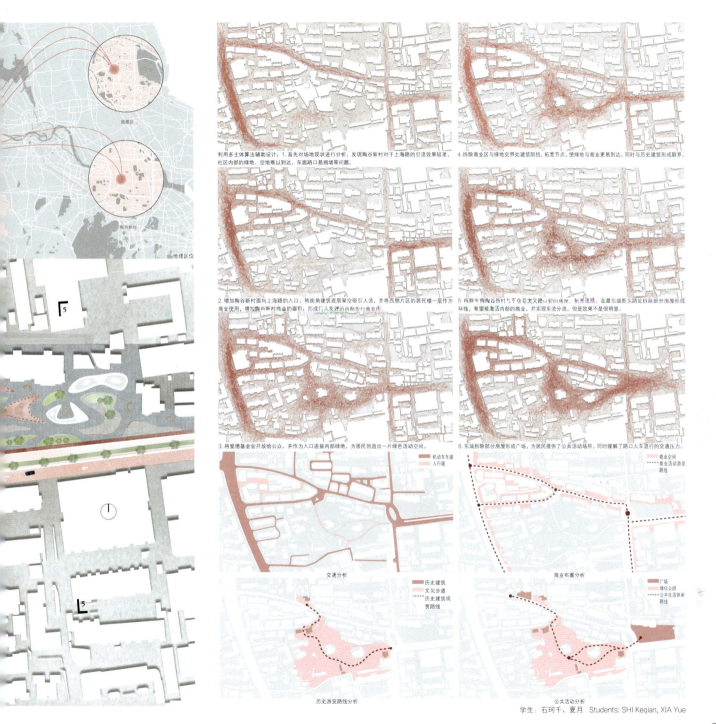

游牧木构——基于轻型可变木结构的可移动帐幕类景观装置设计与建造
NOMADIC WOOD STRUCTURE—DESIGN AND CONSTRUCTION OF MOVABLE TENT-TYPE LANDSCAPE INSTALLATIONS BASED ON LIGHTWEIGHT VARIABLE WOOD STRUCTURES

梁宇舒 缪晓东
LIANG Yushu MIAO Xiaodong

教学目标

　　世界风土建筑中的木构建筑类型呈现出较为丰富的研究成果，而蒙古包的材料以木骨架为主，覆盖以毛毡，却鲜有学者关注其建构层面的技术贡献，相关研究大多停留在文化象征层面。国内利用蒙古包进行的新结构体系研究，则大多以现代材料替换原材料，形式的延续较多而结构逻辑的延续与创新少。

　　本研究型设计题目试图从材料性能、节点构造、制作工艺、搭建过程、结构受力分析等层面，重现蒙古包建筑作为游牧建筑典范之建构逻辑。设计计划从节点研究、网架结构、动态收合、加工工艺、实地搭建、功能需求等入手，设计并实现一座帐幕类景观装置。课程成果需对该设计开展的建造实验进行详细的介绍，最终总结出此次搭建实验对未来变木结构装置发展和应用的经验和启示。

题目简述

　　木材是国际公认的最具有可持续设计特征的建筑材料之一，其材质、性能等特点均与"双碳"目标的要求相契合。如何在乡村振兴的大背景下，依托中国本土特色，应用低碳结构体系和建筑材料，是建筑师通过其绿色建筑设计及创意响应"双碳"目标要求需要解决的重要课题。本课题以蒙古包及其建筑技艺所反映的帐幕类建筑为研究对象，结合文献研究和地方技艺访谈，通过在木工坊的基础木工技能学习与训练，要求学生为短期游牧的牧民在自然生态区域内设计一处用地面积2 m×2 m、遮盖面积为4 m²左右的建筑空间，以满足停留、休憩、居住的功能需求。课题通过木构实践不断探索问题，用设计的方法研究和解决问题，最终通过建构作品的完成与落地来实现研究成果的社会价值。

Teaching objectives

　　Wooden structure building types in the world's vernacular architecture have shown relatively rich research results. But the material of the Mongolian yurt is mainly a wooden frame covered with felt. However, few scholars have paid attention to their technical contributions at the construction level, and most of them have remained at the research level of cultural symbols. The research on new structural systems using Mongolian yurts in China mostly replaces raw materials with modern materials. There is much continuation in form, but little continuation and innovation in structural logic.

　　This research-oriented design project attempts to reproduce the construction logic of Mongolian yurt architecture as a model of nomadic architecture from aspects such as material properties, node structure, production technology, construction technology, and structural stress analysis. The design plan starts from node research, grid structure, dynamic folding, processing technology, on-site construction, functional requirements, etc., to design and implement a tent-type landscape installation. The course results require a detailed account of the construction experiments performed in this design. Finally, the experience of this construction experiment and the inspiration for the future development and application of variable wood structure devices are summarized.

Brief description of the topic

　　Wood is internationally recognized as one of the most sustainable building materials with sustainable design features. Its material, performance and other characteristics all meet the requirements of the carbon peaking and carbon neutrality goals. How to adopt low-carbon structural systems and building materials based on China's local characteristics in the context of rural revitalization is an important issue that architects need to solve through green building designs and creativity to meet the carbon peaking and carbon neutrality goals. This topic takes the Mongolian yurts and the tent-style architecture embodied in its construction technology as the research object. It combines documentary research and interviews on local craftsmen. By learning and training basic carpentry skills in a wood workshop, students are required to design a building space for short-term nomadic herders in natural ecological areas. The internal design area is 2 m × 2 m and the building space covers an area of about 4 m², which meets the functional needs of staying, resting and living. This topic continuously explores problems through the practice of wooden structures, uses design methods to research and solve problems, and finally realizes the social value of research results through the realization and implementation of structural works.

网架结构

1. 梯形木框组合
2. 底座木框组合
3. 折叠木地板（非必需）
4. 剪刀型单元
5. 金属合页节点 1
6. 剪刀型单元
7. 滑动轴组合 1
8. 滑动轴—滑槽
9. 滑动轴组合 2
10. 木框组合 1
11. 木框组合 2
12. 剪刀型单元—子母铆钉
13. 金属合页节点 2

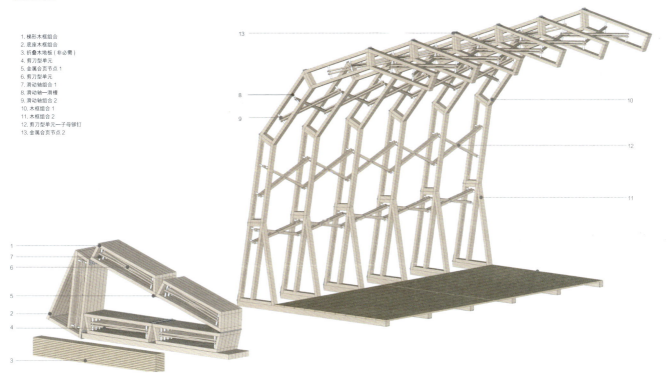

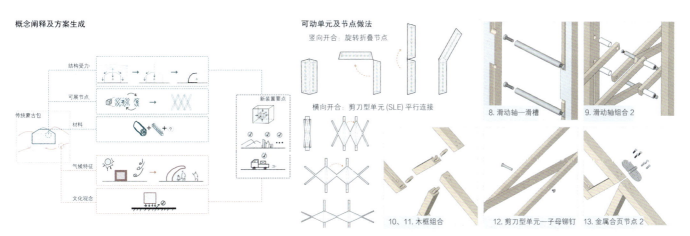

101

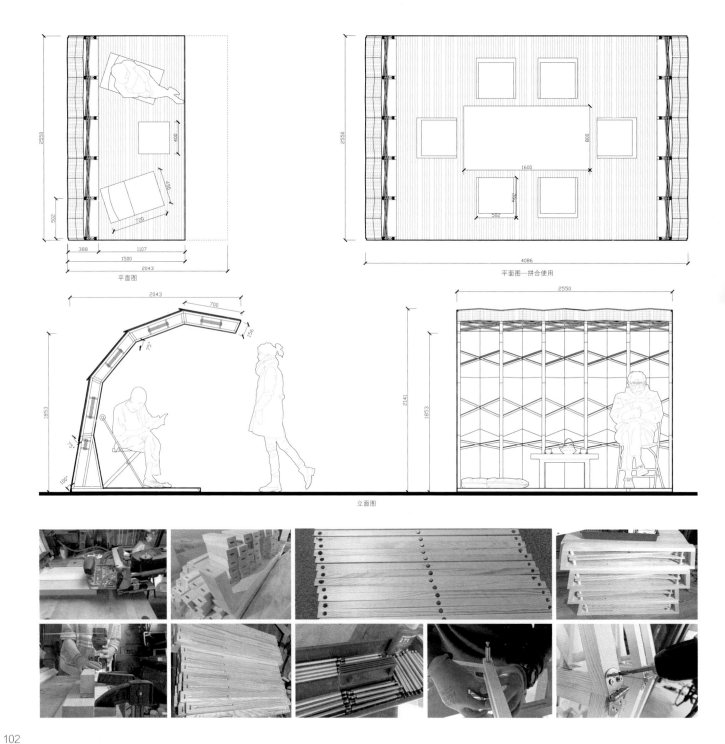

平面图

平面图—拼合使用

立面图

金属合页节点

滑动槽

滑动轴组合

剪刀型单元—子母铆钉

搬运登场地

竖向展开1

竖向展开2

竖向展开3

铺设毛毡

学生：高禾雨　Student: GAO Heyu

新城生活服务基础设施设计研究
RESEARCH ON THE DESIGN OF NEW TOWN LIFE SERVICE INFRASTRUCTURE

王 铠
WANG Kai

项目简述

新城区是中国城市化建设的重要内容，在过去的20年一直是中国建筑行业面临的重要工作领域。大量的住宅建设和城市新民居，以及与之配套建设的大量公共基础设施（体育、文化、会展、商业服务、交通设施），往往在激进的城市扩张背景下过度建设而后期运营艰难，也未能发挥激活社区活力的职能。在未来的10—20年间，新城公共基础设施的生活服务功能，关乎城市居民生活的持续高质量提升与城市的持续活力，将会继续成为建筑业面临的重要议题之一。

项目选取某经济发达地区城市新城的混合城市用地区域公共基础设施作为研究对象，其用地规模为3—5 hm²，建筑面积为2万—4万 m²，进行生活服务基础设施设计研究。内容包括修建性详细规划和建筑单体设计。尽管新城建设的巨大成就毋庸置疑，然而其追求速度、效率的同时，无可否认地忽视了整体环境人性化、归属感、公共性的社区意义，本设计选题正是希望对此做出回应。

重点问题

1) 新城开发建设模式；
2) 现行法规及技术体系；
3) 新城公共基础设施的类型与模式；
4) 产业、业态与空间；
5) 行为、体验与空间；
6) 城市空间可持续发展。

工作方式

分为五个阶段：其一，研究项目背景，分析规划条件和客户/市场；其二，做项目策划及研究案例（功能类型、规划布局模式案例）；其三，综合研究街区形态、土地利用、空间选型与组合、生成概念规划；其四，深化完成规划设计方案、典型建筑设计方案及其技术节点的深化；其五，通过几个阶段的学习研究，最终完成毕业设计。

Project introduction

New urban areas are an important part of China's urbanization construction and have been an important work area for the Chinese construction industry in the past two decades. A large number of residential construction and new urban residents, as well as a large number of supporting public infrastructures (sports, culture, exhibitions, commercial services, transportation facilities), are often over-constructed in the context of radical urban expansion, making it difficult to operate in the later period and failing to fulfill the function activating the vitality of the community. In the next 10-20 years, the life service function of new city public infrastructure which are related to the continued high-quality improvement of urban residents' lives and the continued vitality of the city, will continue to become one of the important issues faced by the construction industry.

The project selects the public infrastructure in a mixed urban land area of a new city in an economically developed area as the research object, with a land size of 3-5 hm² and a building area of 20,000-40,000 m², to research on the design of living service infrastructure. The content includes architectural detailed planning and individual building design. Although there is no doubt about the great achievements of the new city construction, while pursuing speed and efficiency, it undeniably ignores the significance of community in terms of humanization, a sense of belonging and publicity of the overall environment. This design topic hopes to respond to this.

Key issues

1) Development and construction models of new urban areas;
2) Current regulations and technical systems;
3) Types and models of public infrastructure in new urban areas;
4) Industries, business forms and space;
5) Behavior, experience and space;
6) Sustainable development of urban space.

Working methods

It is divided into five stages: first, research the project background, analyze the planning conditions and customer/market; second, do the project planning and study the cases (functional types, planning layout model cases); third, conduct comprehensive research on the block forms, land use, space selection, and generate conceptual planning; fourth, deepen the completion of planning and design plans, typical architectural design plans and their technical nodes; fifth, through several stages of study and research, complete the graduation project.

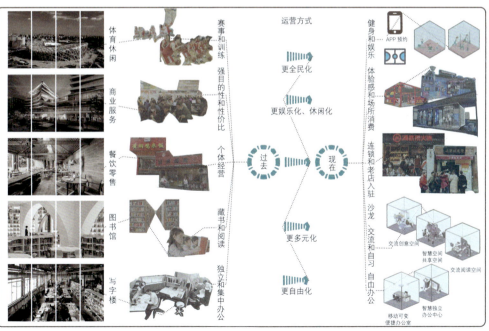

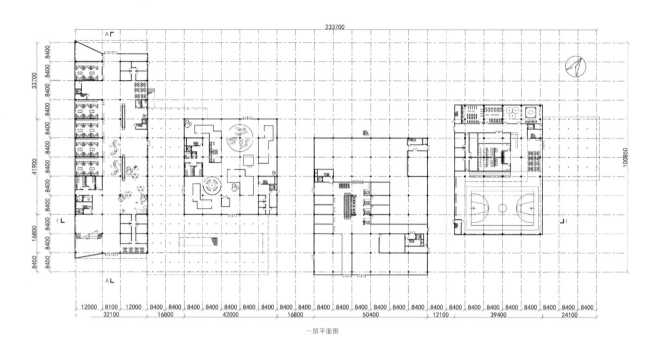

一层平面图

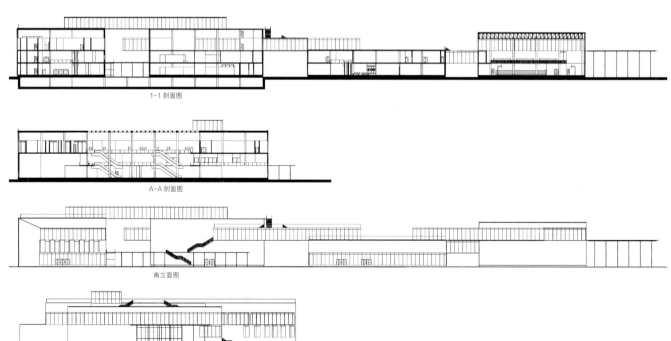

1-1 剖面图

A-A 剖面图

南立面图

东立面图

学生：唐诗诗　Student: TANG Shishi

基于复杂地形的参数化景观建筑设计
PARAMETRIC LANDSCAPE ARCHITECTURAL DESIGN BASED ON COMPLEX TERRAIN

尹航
YIN Hang

总体介绍

本次毕业设计的主题主要是基于一定复杂地形的景观建筑设计。其设计内容涉及景观场地设计和基于实际旅游、运动、接待需求的功能性建筑设计。

设计场地将在两块实际场地中进行选择,其中一处是滨水山林地,另一处是一座山谷村落。可根据实地考察情况选取。

本次毕业设计希望在地形分析、场地整理和建筑布局等各阶段适当加入参数化的设计路径,在设计前期对场地平面、建筑体块和景观流线布局的设计机制进行一定研究。

地块情况

本次设计提供两块实际场地供选择,其一位于江苏省盱眙县的铁山寺风景区附近,这里山水资源丰富,临近天泉湖水库,总体丘陵面积超过 2 km²,但设计场地限定约为 50 hm²。设计要求是在场地内结合地形与周边限制,设计合理的景观分区、道路系统和户外活动场地等,在其中布置总面积不超过 10 000 m² 的度假、接待、运动和康养功能建筑若干。

其二位于黄连村——安徽省休宁县山区的一座小型村庄。黄连村位于皖、赣、浙三省交界,是原徽饶古道上的一座历史村落,有 30 余户人家,四面环山,可用规划面积约为 20 hm²。拟将当地民房进行流转,设计一处生态康养型度假村,建筑体量在 5 000—10 000 m²,接待人员量不超过 200 人。本设计是城乡振兴工作营设计的延伸,可与工作营 2022 年在附近的梓坞村已经完成的旅游线路分析与概念设计内容进行结合。

在实地调研、选定地块后,同学们将联合完成环境分析、参数化工具的实验与运用,总体场地布局设计、建筑概念设计等。之后,根据总体设计情况,组内同学各自再选择一处建筑或场地景观进行深化设计。

General introduction

The theme of this graduation project is the landscape architectural design based on certain complex terrain. The design content involves landscape site design and functional architectural design based on actual tourism, sports, and reception needs.

The design site will be chosen between two actual sites, one of which is a waterside mountain woodland and the other is a valley village. The selection can be made based on the on-site inspections.

This graduation project hopes to appropriately incorporate parametric design paths into various stages such as terrain analysis, site arrangement, and building layout, and conduct certain research on the design mechanism of site planes, building blocks, and landscape flowline layout in the early stages of design.

Site condition

This design provides two actual site options for selection. One is located near the Tieshan Temple Scenic Area in Xuyi County, Jiangsu Province. It is rich in mountain resources and is close to the Tianquan Lake Reservoir. The overall hilly area exceeds 2 km², but the design site is limited to about 50 hm². The design requirement is to combine the terrain and surrounding restrictions within the site, and design reasonable landscape zoning, road systems and outdoor activity venues, etc., and arrange several functional buildings for vacation, reception, sports and health care with a total area of no more than 10,000 m².

The second is located in Huanglian Village—a small village in Xiuning Mountain, Anhui Province. The village is located at the junction of Anhui, Jiangxi and Zhejiang Provinces. It is a historical village on the original Huirao Ancient Road. It has more than 30 households and is surrounded by mountains. The usable planning area is about 20 hm². It is planned to transfer local houses and design an ecological health resort. The building volume is around 5,000–10,000 m², and the number of reception staff does not exceed 200 people. This design is an extension of the design done by the Urban-Rural Revitalization Work Camp. The work camp have completed tourism route analysis and conceptual design content in nearby Ziwu Village in 2022, which can be combined with this design.

After on-site investigation and land selection, students will jointly complete environmental analysis, experimentation and application of parametric tools, overall site layout design, architectural concept design, etc. After that, based on the overall design situation, students in the group can each choose a building or a site landscape for further in-depth design.

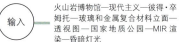

| 输入 | 火山岩博物馆—现代主义—彼得·卒姆托—玻璃和金属复合材料立面—透视图—国家地质公园—MIR渲染—昏暗灯光 | 博物馆—现代主义—比雅克·英格斯—复合材料立面、内部红灯和金属结构—人视点—火山和森林—MIR渲染—自然光 | 博物馆大厅内部—现代主义—比雅克·英格斯—玻璃和石材复合材料立面、内部红光和金属结构—人视点—森林公园—MIR渲染—自然光 | 博物馆、L形公共建筑—现代主义—比雅克·英格斯—玻璃和金属复合材料结构—鸟瞰图—火山和森林以及小径—MIR渲染—自然光 |

输出

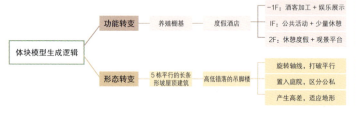

将建筑放置在大草坪上—人们步行—树木和石头—国家森林公园、美丽的花园—倾斜的屋顶—轻微的地形起伏—MIR渲染—自然光

度假酒店—条形建筑—山地建筑—斜屋顶建筑—大草坪—树木和石头—火山岩—步行道—逼真的渲染—国家森林公园、美丽的风景—自然采光

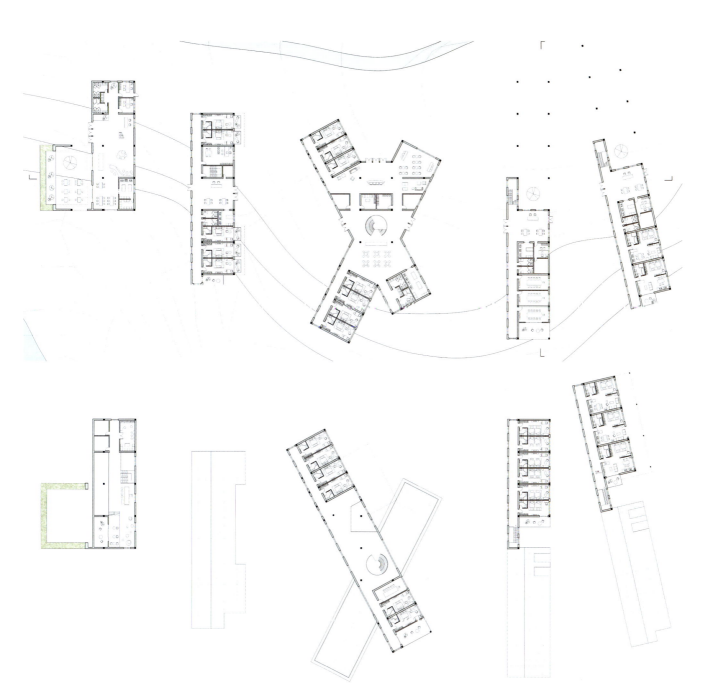

学生：夏月　Student: XIA Yue

建筑设计研究（一）ARCHITECTURAL DESIGN RESEARCH 1

基本设计
BASIC DESIGN

傅 筱
FU Xiao

研究课题
    宅基地住宅设计

教学目标
    课程从"场地、功能、行为、结构、经济性"等建筑的基本问题出发，通过宅基地住宅设计训练学生对建筑逻辑性的认知，并让学生理解有品质的建筑是以基本问题为基础的整合设计。

设计内容
    1）每个设计工作组由2—3位同学组成，训练学生设计合作。
    2）在A、B两块宅基地内任选一块进行住宅设计。

设计要求
    1）基地位于南京郊区某村。使用者设定为一对夫妇，他们有一个4岁的儿子。住宅兼做丈夫的个人独立工作室，妻子为全职太太。工作室性质由设计者自定。
    2）建筑用地边界不得超过原有宅基地范围（出入口踏步除外），边界位置也不得随便移动。一层悬挑的雨篷、二层悬挑的阳台（凸窗）可适当超出用地边界。
    3）建筑层数2层，室内外高差0.45 m。
    4）室外停小汽车2—3辆（不考虑室内停车）。
    5）须进行场地布置设计。

设计成果
    1）完成平、立、剖面图纸，平面图比例为1:50，剖面图、立面图比例为1:100，剖面图要求布置家具以及表达人的行为。
    2）表达空间关系的三维剖透视图1—2个，比例不低于1:50，要求材料填充，必须有家具和人的行为表达。
    3）比例不低于1:20的表达设计意图的外墙大样图，不少于2幅。
    4）有利于表达形体和空间的透视图，可以是渲染图，也可以是模型照片。
    5）其他有利于设计意图表达的图纸。

Research topic
    Homestead residential design

Teaching objectives
    The course starts from the basic issues of architecture such as "site, function, behavior, structure, and economy". Through homestead residential design, it

trains students to understand the logic of architecture and allows students to understand that quality architecture is based on basic issues.

Design content
1) Each design working group consists of 2-3 students to train students in design collaboration.
2) Choose either of the two plots A and B within the homestead for residential design.

Design requirements
1) The base is located in a village in the suburbs of Nanjing. The users are configured as a couple with a 4-year-old son. The house also serves as the husband's personal independent studio, and the wife is a full-time housewife. The nature of the studio is determined by the designer.
2) The boundary of the building land shall not exceed the scope of the original homestead (except for the entrance and exit steps), and the boundary position shall not be moved at will. The cantilevered awning on the first floor and the cantilevered balcony (bay window) on the second floor can appropriately exceed the land boundary.
3) The building has 2 floors and the indoor-outdoor height difference is 0.45 m.
4) Park 2-3 cars outdoors (indoor parking is not considered).
5) Site layout design is required.

Design results
1) Complete the floor plan, elevation and section drawings with a scale of 1:50 for the floor plan, 1:100 for the section and elevation. The section requires the arrangement of furniture and the expression of human behavior.
2) 1-2 three-dimensional cross-sections that express spatial relationships, with a scale of no less than 1:50. Material filling is required, and furniture and human behavior must be expressed.
3) No less than 2 exterior wall detail drawings expressing the design intention with a scale of not less than 1:20.
4) Perspective drawings that are conducive to expressing form and space, which can be renderings or model photos.
5) Other drawings that are conducive to the expression of design intentions.

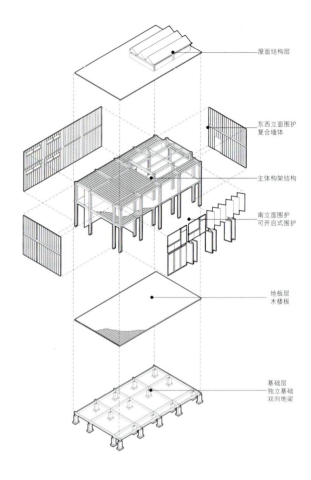

屋面结构层

东西立面围护复合墙体

主体构架结构

南立面围护可开启式围护

地板层木楼板

基础层独立基础双向地梁

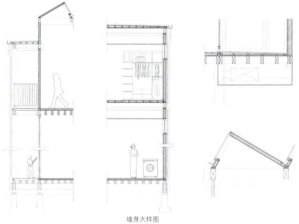

墙身大样图

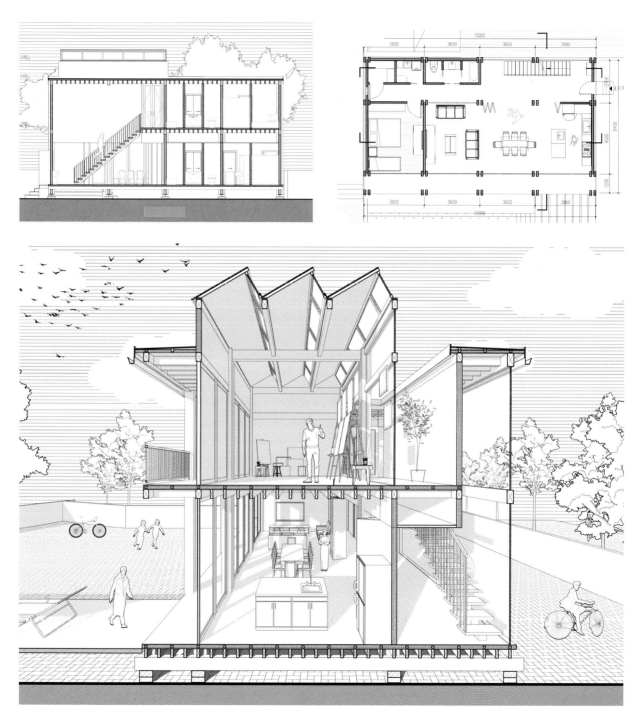

学生：陆禹名，刘传，姜辰达　Students: LU Yuming, LIU Chuan, JIANG Chenda

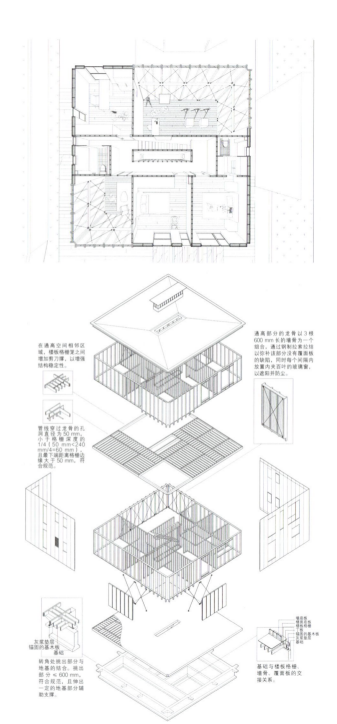

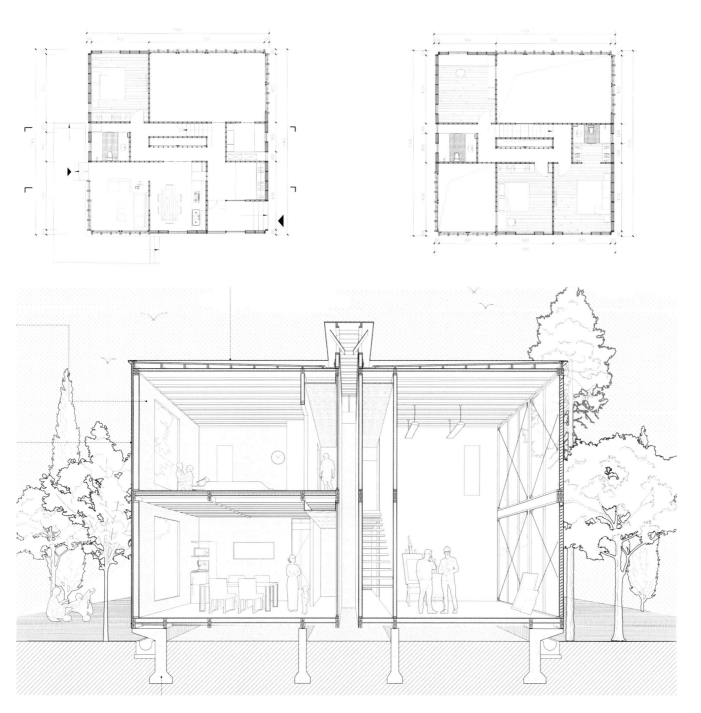

学生：赵潇艺，张桂煜，陈哲 Students: ZHAO Xiaoyi, ZHANG Guiyu, CHEN Zhe

建筑设计研究（一）ARCHITECTURAL DESIGN RESEARCH 1

基本设计
BASIC DESIGN
冷 天
LENG Tian

教学目标

当下中国的城市建设发展，业已从"增量开发"走向"存量更新"。其中，文物建筑、历史建筑、文化遗产等大量历史性建筑，面临着如何在保存特有历史文化价值的前提下，充分活化利用其原有空间（内部、外部）的难题。本课程希望引导学生通过一个真实的案例，直面上述的双重矛盾，理解设计的逻辑性与综合性，在历史和现实之间取得平衡，并通过一个复有创造性的设计来激发空间的活力。

设计内容

1）首都水厂清凉山水库旧址

首都水厂清凉山水库旧址位于南京市鼓楼区清凉山，是中国第一座自主建设的现代化自来水厂——南京首都水厂自来水系统中不可或缺的一环。清凉山水库充分利用清凉山的地形标高，在用水低谷时进行蓄水，在用水高峰时向外供水，用于调剂市内用水，补救首都水厂供水不足引起的局部缺水，以及遇有火灾额外增加的需水量，对当时全市供水发挥了直供和补缺的作用。其地下主体结构为钢筋混凝土框架剪力墙结构，建筑面积约为 2 600 m²，地下蓄水池容积约为 1 万 m³。

问题与设计任务：地下蓄水池主体承重结构经常年水体腐蚀受到了不同程度的损伤与破坏，须在适宜修缮手段的基础上，对水库蓄水池闲置的空间进行活化利用，挖掘并显现其独特的历史价值和地位。

2）东南楼

东南楼位于南京大学鼓楼校区北园东侧，临近天津路，为一栋三层砖木混合结构建筑。建筑采用传统歇山顶，平面为"工"字形，坐东朝西，与西侧西南楼共同组成以教学楼为中心的南京大学鼓楼校区主轴线空间。建筑总面积约为 7 000 m²。

问题与设计任务：东南楼将成为南京大学医学院新的公共教学空间，设计须在不破坏其历史建筑外貌的条件下提升空间品质，充分利用闲置的屋架层空间，彰显其历史文化的价值。

设计范围和研究范围

各个对象的设计范围为：

1）清凉山水库地上与地下部分，须同时考虑主体结构的修缮手段与功能重置以激活空间。

2）东南楼三层与屋架层空间，须考虑如何有效使用屋架层空间。

除设计范围外，各组可自行设定扩大的研究范围，研究内容可以是对历史、景观、交通、功能、视线等问题的分析。

Teaching objectives

At present, urban construction and development in China have shifted from "incremental development" to "stock renewal". Among them, a large number of historical buildings such as cultural relic buildings, historical buildings, cultural heritage, etc., are faced with the problem of how to fully activate and utilize their original spaces (internal and external) while preserving their unique historical and cultural value. This course aims to guide students through a real case to confront the above-mentioned dual contradictions, to understand the logic and comprehensiveness of design, to strike a balance between history and reality, and to stimulate the vitality of space through a creative design.

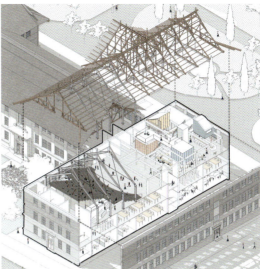

Design content

1) The former site of Qingliangshan Reservoir of Capital Water Plant

Located in Qingliangshan, Gulou District, Nanjing City, Qingliangshan Reservoir of Capital Water Plant is China's first independently constructed modern water plant—an indispensable link in the water system of Nanjing Capital Water Plant. The Qingliangshan Reservoir makes full use of the topographic elevation of Qingliangshan to store water when water use is low and supply water to the outside during peak water use. It is used to adjust the city's water supply, remedy local water shortages caused by insufficient water supply from the Capital Water Plant, and meet the additional water demand in case of fires, playing a direct supply and supplementary role in the city's water supply at that time. Its main underground structure is a reinforced concrete frame shear wall structure, with a construction area of about 2,600 m². The underground reservoir has a volume of about 10,000 m³.

Problems and design tasks: The main load-bearing structure of the underground reservoir has suffered varying degrees of damage and destruction caused by long-term water body corrosion. It is necessary to activate and utilize the idle space of the reservoir on the basis of suitable repair methods, excavate and reveal its unique historical value and status.

2) Southeast Building

Located on the east side of the North Park of Nanjing University's Gulou Campus, close to Tianjin Road, Southeast Building is a three-story brick and wood mixed-structure building. The building adopts a traditional hip roof and a planar layout in the shape of a "Gong" character, facing east and sitting west. Together with the Southwest Building on the west side, it forms the main axis space of the Nanjing University's Gulou Campus with the teaching building as the center. The total construction area is approximately 7,000 m².

Problems and design tasks: Southeast Building will become a new public teaching space for the Medical School of Nanjing University, and the design is necessary to improve the quality of the space without destroying the appearance of its historical building, make full use of the idle attic space and highlight its historical and cultural value.

Design scope and research scope

The design scope of each object is:

1) For the above-ground and underground parts of Qingliangshan Reservoir, it is necessary to consider both the repair methods of the main structure and the reset of functional to activate the space.

2) For the space on the third floor of Southeast Building and the idle attic space, it is necessary to consider how to effectively use the idle attic space.

In addition to the design scope, each group can set its own expanded research scope. The research content can include analysis of history, landscape, transportation, function, sightline and other issues.

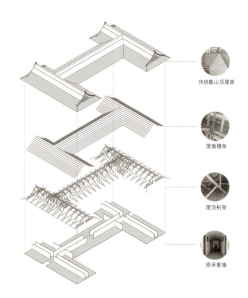

传统歇山顶屋面
屋面檩条
屋顶桁架
原承重墙

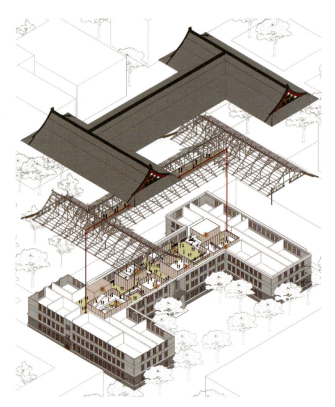

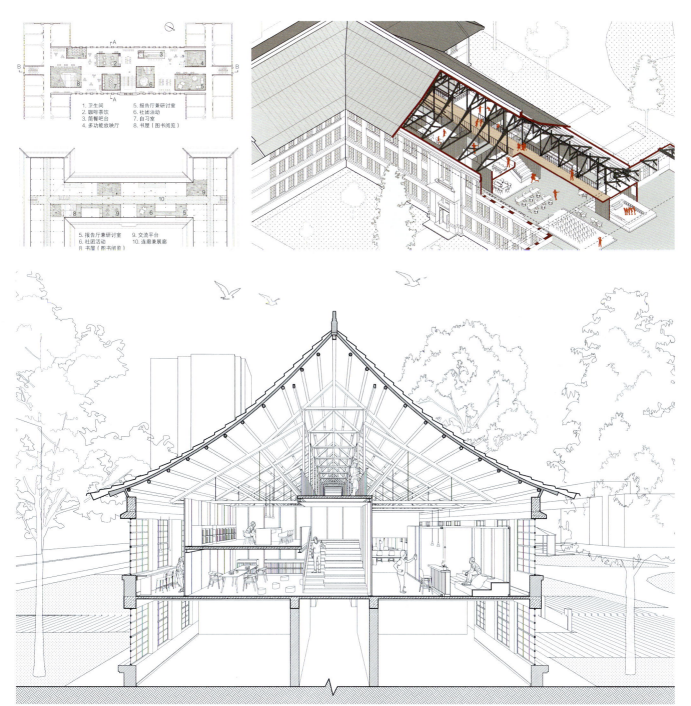

学生：何佳慧，郭辉，张金库 Students: HE Jiahui, GUO Hui, ZHANG Jinku

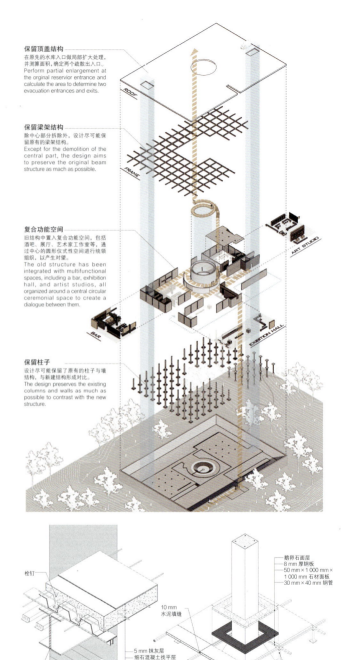

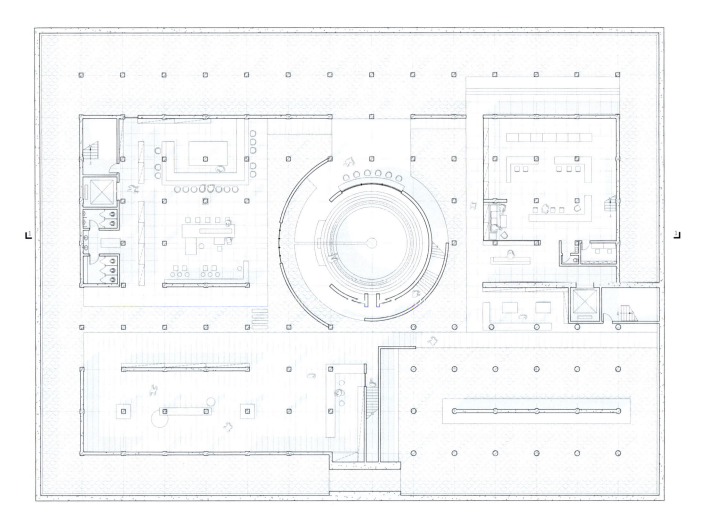
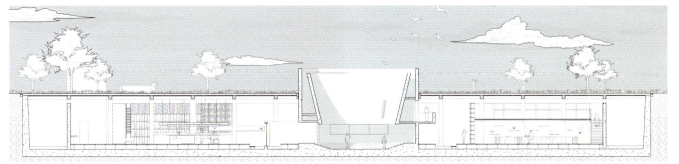

学生：黄佳怡，陈晨，左琪 Students: HUANG Jiayi, CHEN Chen, ZUO Qi

建筑设计研究（一）ARCHITECTURAL DESIGN RESEARCH 1

基本设计
BASIC DESIGN

潘幼建　万军杰
PAN Youjian　WAN Junjie

教学目标

结构常常只被当作建筑克服重力的工具，其在建筑学范畴上具有的意义常常被忽视了，我们也常在设计的最后阶段才展开对于结构的讨论，而忽视结构在设计的最初阶段对概念的推动作用。一种合适的结构形式可以推动一座优秀建筑的产生，结构的创新也是建筑创新的重要推动力。

本次教学的目的是训练建筑设计的概念意识，并将概念贯穿到形式、空间、结构、细部的设计当中，以生成呈现逻辑严密、设计意图明确的建筑。

结构可以成为实现建筑概念和空间目标的手段，从设计的最初阶段就有结构的思考，建立起结构对空间感知影响的敏锐度。

概念与结构的设计训练的主要目的是训练建筑设计的"底层逻辑"，即对建筑的基本认知，以及整合的方法和技巧，对形式的敏感性和把控力。

本次教学强调概念与结构的研究议题。研究型设计教学不对建筑规模和类型做具体的要求，而希望学生能够通过观察和留意自己生活的环境，建立发现问题并用建筑去解决问题的意识。这是对一种设计方法的训练，其结果是具有开放性和多样性的。

本次教学训练学生关注建筑最基本的要素，从基本要素出发去寻找新的突破——一个从环境到概念和结构的连贯的构思过程。

设计任务

设计课程从教师对案例的分享开始，以建立学生对概念与结构的基础意识。学生观察和调研自己所在的南京大学浦口校区，从几个备选场地中挑选一个，发现问题，依据功能主题拟定任务书，提出空间概念和结构策略，回应场地，应对问题。

结构形式来源于学生对于案例的分析（可与教师分享案例有重叠），通过分析理解结构原理，及结构与概念和空间的整体关系，以案例为基础做概念和结构的抽象和延伸，将其运用到自己的设计当中。

规模：不对建筑规模做具体的要求。但要注重建筑与周边环境的关系。

功能：依据功能主题，基于场地和空间规划自主确定项目功能构成及各部分面积，功能宜复合且具有一定的差异性。

设计成果

建筑总图、平面图、立面图、剖面图，白底黑色线稿图，两张 A1 图版。具体比例依据建筑规模调整。

表达建筑概念的效果图 6 张左右，每张效果图满铺 A1 图版。

建筑结构概念模型，具体比例依据建筑规模调整。

Teaching objectives

Structure is often only used as a tool for architecture to overcome gravity, and its significance in the field of architecture is ignored. We often discuss structure in the final stages of design, neglecting the role that structure can play in driving concepts from the very beginning. A suitable structural form can promote the production of an excellent building, and structural innovation is also an important driving force for architectural innovation.

The purpose of this teaching is to train the conceptual awareness of architectural design and integrate the concepts into the design of form, space, structure and details,

in order to present a building with strict logic and clear design intention.

Structure can be a means to achieve architectural concepts and spatial goals. From the initial stage of design, there should be a consideration of the structure, and a sensitivity to the impact of structure on spatial perception should be established.

The main purpose of conceptual and structural design training is to train the "underlying logic" of architectural design, that is, the basic understanding of architecture, integrated methods and techniques, and sensitivity and control of form.

This teaching emphasizes the research topics of concepts and structures. The research-based design teaching does not set specific requirements for the scale and type of buildings, but hopes that students can observe and pay attention to the environment in which they live, establish an awareness of identifying problems and using architecture to solve them. It is the training of a design method whose results are open-ended and diverse.

This teaching trains students to pay attention to the most basic elements of architecture, and start from the basic elements to find new breakthroughs—a coherent ideation process from environment to concept and structure.

Design tasks

The design course starts with teachers sharing cases to build students' basic awareness of concepts and structures. Students observed and investigated the Pukou Campus of Nanjing University where they were located, selected one of several alternative sites, discovered problems, formulated a task statement based on functional themes, proposed spatial concepts and structural strategies, responded to the site, and responded to the problems.

The structural form comes from students' analysis of case studies (which may overlap with the cases shared by the instructor). Through analysis, they understand the structural principles and the overall relationship between structure, concept and space, abstract and extend concepts and structures based on case studies and apply them to their own designs.

Scale: There are no specific requirements for building scale. But we should pay attention to the relationship between the building and the surrounding environment.

Function: According to the functional theme, the functional composition of the project and the area of each part are determined independently based on the site and space planning. The functions should be composite and have certain differences.

Design results

Architectural general plan, floor plans, elevations, sections, black line drawings on white background, two A1 plates. The specific proportion should be adjusted according to the scale of the building.
There are about 6 renderings expressing the architectural concept, and each rendering covers the entire A1 plate.
Concept model of the building structure, the specific proportions are adjusted according to the scale of the building.

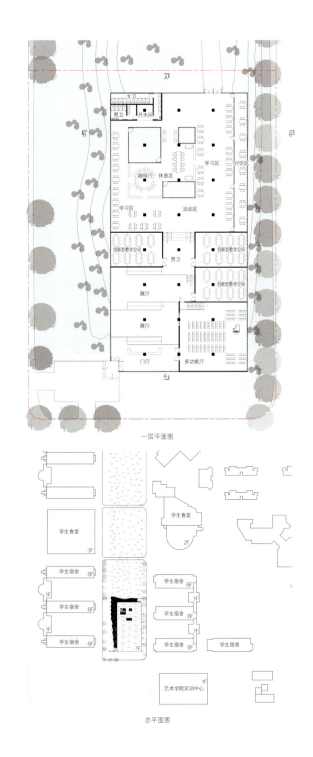

一层平面图

总平面图

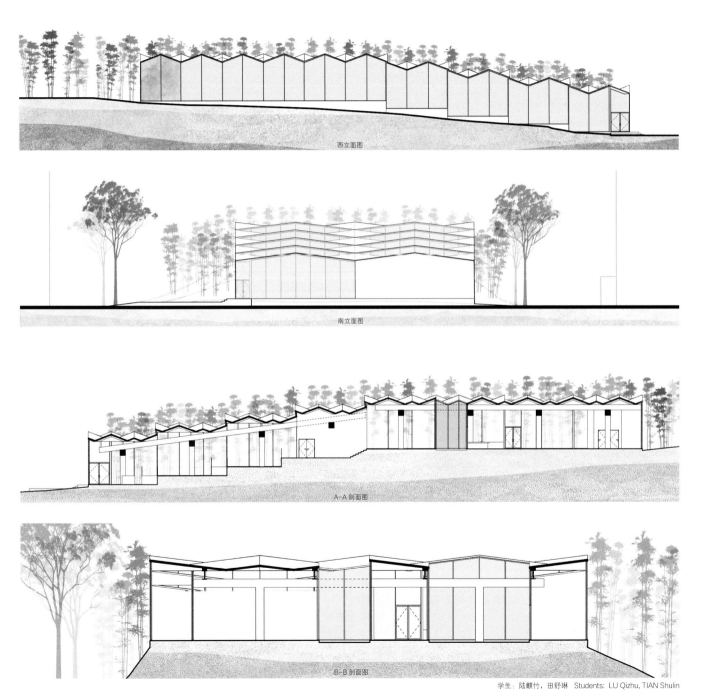

学生：陆麒竹，田舒琳　Students: LU Qizhu, TIAN Shulin

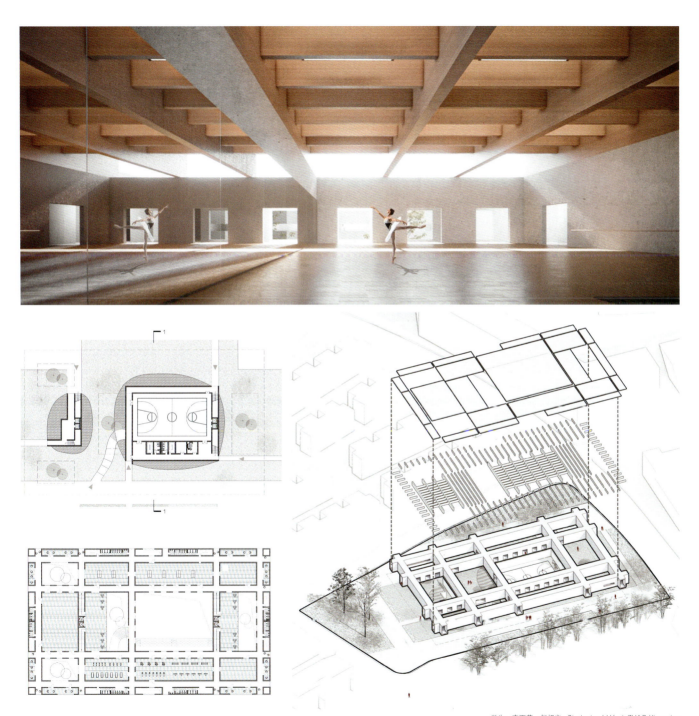

学生：李雨茜，赵相宁 Students: LI Yuxi, ZHAO Xiangning

建筑设计研究（一）ARCHITECTURAL DESIGN RESEARCH 1

概念设计
CONCEPTUAL DESIGN

鲁安东
Lu Andong

课程内容

本学期概念设计课程以 2022 深圳双年展为背景，聚焦"生息"这一概念，以新的认知起点重新审视建筑学科的相关性，并对急迫的城市问题做出正面回应。生息是共生的价值观，也是绵延的规律性，同时承载着共生的价值观和延续性的法则。2022 深圳双年展以"城市生息"为主题；"城市生息"秉持的是多样的自然要素的和谐共生、生生不息；"城市生息"追求的是通过对生活方式的塑造，实现人类繁衍生息；"城市生息"探索的是循环与平衡的发展之道，通过城市创新与人的行动，共同走向未来。这三个维度分别体现了自然的价值、城市的价值和人的价值。

高密度条件下的生态共生、"超地方的"和"泛人类的"流动交换，以及技术带来的全新的增强环境的可能性，共同构成了当代新型人类聚居的特征。当前城市发展面临着大转型，无论是从低碳可持续角度，还是从存量更新或者收缩城市的角度，都要求我们转向以生命为中心，更加精明、适变和包容的设计。

学生将二人一组，首先构建一种"生息"类型，为其建立概念图谱，并且找到其与当代人文、技术的关联，进而加以应用，并设计一个原型性的实验方案。课程设计突破传统建筑学的学科边界，在生态哲学、生命科学、数字技术与人文关怀的交叉领域建构动态设计思维。这一概念既指向多物种共栖的生态智慧，也隐喻城市作为有机生命体的新陈代谢规律，更蕴含着技术文明时代人类重新校准与自然关系的伦理自觉。

Course content

This semester's conceptual design course focuses on the concept of "habitat" against the backdrop of 2022 Shenzhen Biennale, re-examining the relevance of the architectural discipline from a new cognitive starting point and responding positively to pressing urban issues. Habitat is the value of symbiosis and the

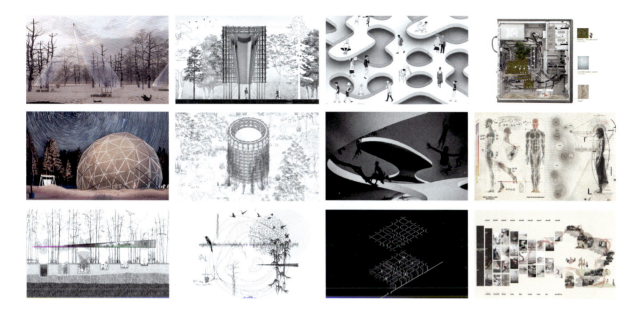

regularity of continuity, carrying both the symbiotic value system and the law of continuity. 2022 Shenzhen Biennale takes "urban habitat" as its theme: "urban habitat" upholds the harmonious coexistence of diverse natural elements and endless growth; "urban habitat" pursues the realisation of the human condition through the shaping of lifestyles; "urban habitat" explores the way of cyclic and balanced development, and moves towards the future together through urban innovation and human action.

The ecological symbiosis under high-density conditions, the "super local" and "pan human" mobility and exchange, as well as the new possibilities of enhancing the environment brought by technology, collectively constitute the characteristics of contemporary new human settlements. Currently, urban development is facing a major transformation, whether from the perspective of low-carbon sustainability, stock renewal or shrinking cities, we are required to shift to life-centred, more astute, adaptable and inclusive design.

Students will work in pairs to first construct a typology of "habitat", establish a conceptual map for it, find its connections with contemporary humanities and technology, and then apply it and design a prototype experimental scheme. The course design breaks through the disciplinary boundaries of traditional architecture and constructs dynamic design thinking at the intersection of ecological philosophy, life sciences, digital technology, and humanistic care. This concept not only refers to the ecological wisdom of multi species coexistence, but also metaphorically represents the metabolic laws of cities as organic life forms, and embodies the ethical consciousness of humans in the era of technological civilization to recalibrate their relationship with nature.

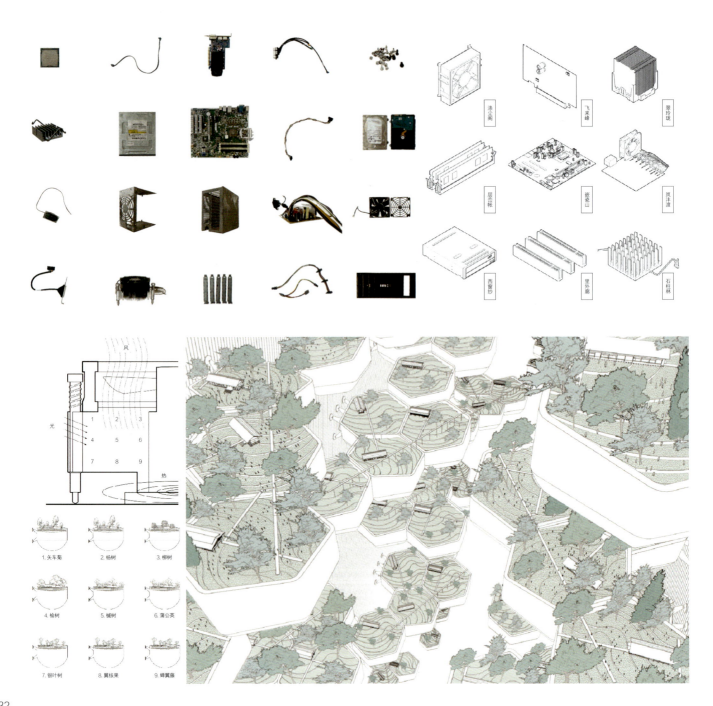

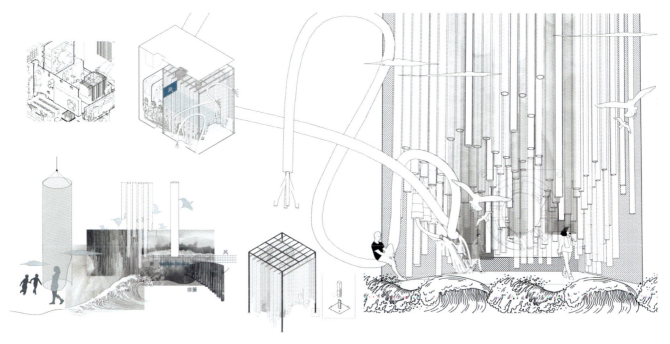
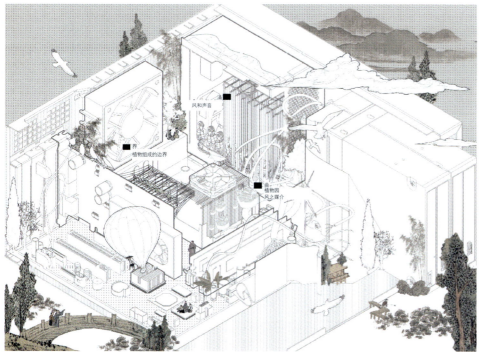
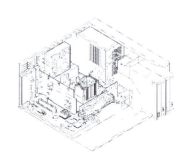

学生：罗宇豪，张百慧，张新雨　Students: LUO Yuhao, ZHANG Baihui, ZHANG Xinyu

学生：徐嘉曼，缪政儒，刘文博睿 Students: XU Jiaman, MIAO Zhengru, LIU Wenborui

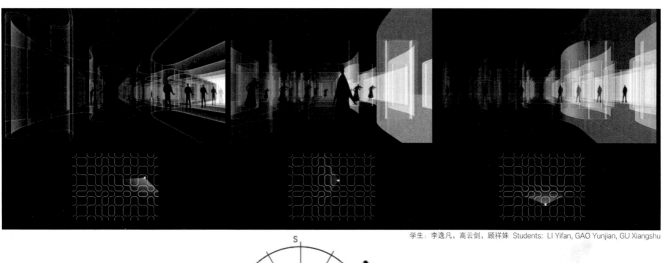

学生：李逸凡，高云剑，顾祥姝 Students: LI Yifan, GAO Yunjian, GU Xiangshu

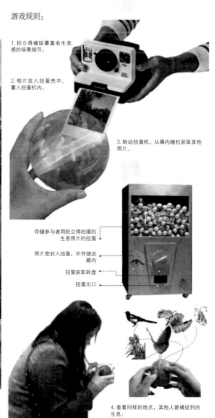

学生：郭烁，崔葆莉 Students: GUO Shuo, CUI Baoli

135

建筑设计研究（一）ARCHITECTURAL DESIGN RESEARCH 1

概念设计
CONCEPTUAL DESIGN
窦平平
DOU Pingping

题目：可感知的共生

课程内容

  本课题探索如何在高度城市化、人工化的环境中与自然要素互惠共生，如何通过营造建筑环境在人类特质与自然要素之间建立隐秘又愉悦的联系。

  人类天生渴望某种与自然世界有关的信息，这是植根于人类基因之中的对自然的依赖性情感反应。与自然世界的联系与大脑的快乐和痛苦中心高度相关。

  共生理论营造了一种动态情境，使得两种或多种异质元素可以在保持张力的状态下共同依存，彼此支撑。

  互利是共生设计的核心目标。

  共生建筑是复合的、生发的，在异质中共存。

  共生建筑是延续的、动态的，在变化中获得稳定。

课程要求

  课程要求提出一种共生建筑的概念，及其相应的空间模式、交互机制、感知方式、关键细节，寻找和呈现环境中隐匿的能量流动与制约关系，探索不以人类为中心的但有益于人类的设计。

Title：Sensible Symbiosis

Course content

  This project explores how to live in symbiosis with natural elements in a highly urbanized and artificial environment, and how to create a hidden yet pleasurable connection between human traits and natural elements through the creation of a built environment.

  Human beings have an innate desire for some kind of information about the natural world, and this is a dependent emotional response to nature that is rooted in our genes. The connection with the natural world is highly correlated with the brain's centers of pleasure and pain.

  Symbiosis theory creates a dynamic situation in which two or more heterogeneous elements can co-depend and support each other while maintaining tension.

  Mutual benefit is the central goal of symbiotic design.

  Symbiotic architecture is composite, generative, and coexists with heterogeneity.

  Symbiotic architecture is continuous and dynamic, and achieves stability through change.

Course requirements

  The course requires the presentation of a concept of symbiotic architecture and its corresponding spatial patterns, mechanisms of interaction, modes of perception, and key details. Seek and present the hidden energy flows and constraints in the environment. Explore designs that are not anthropocentric but beneficial to humans.

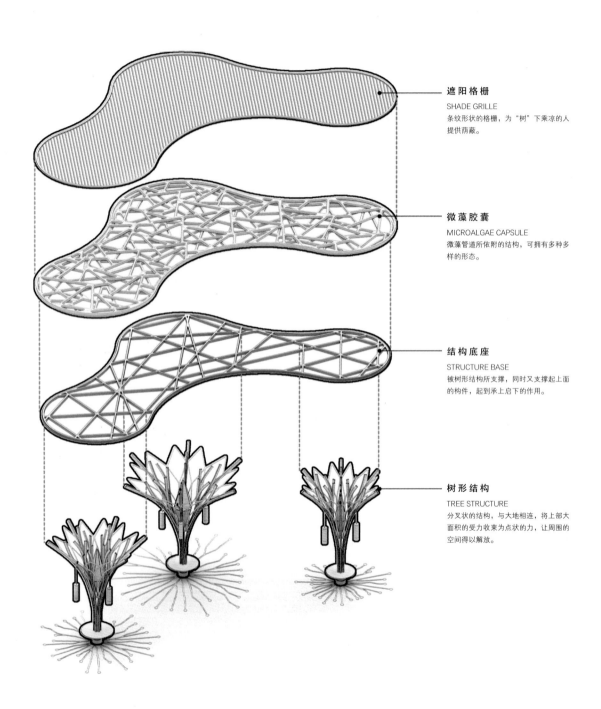

138

学生：文琦，杨兆琪 Students: WEN Qi, YANG Zhaoqi

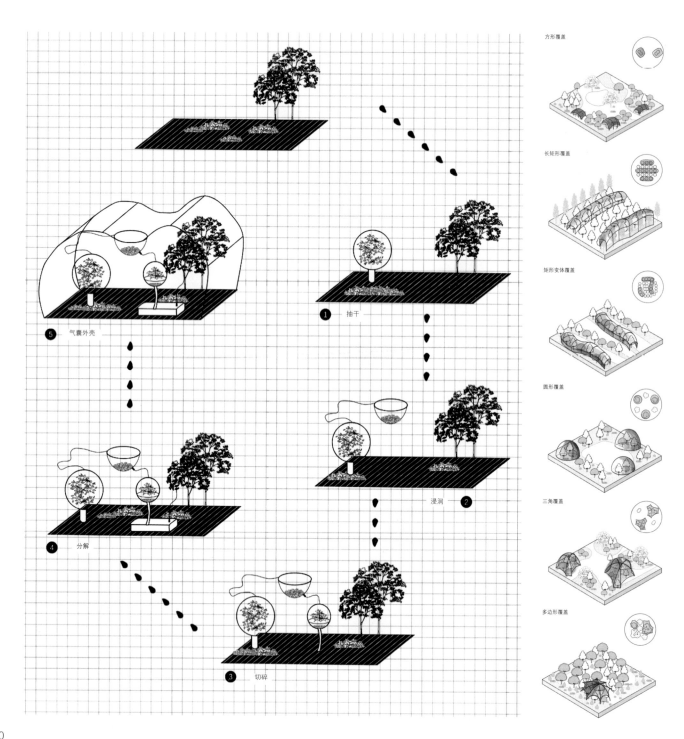

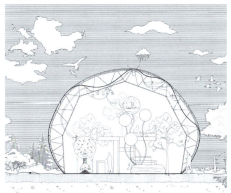
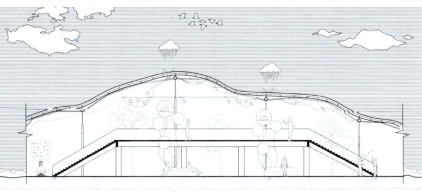

学生：刘未然，徐嘉曼　Students: LIU Weiran, XU Jiaman

建筑设计研究（二）ARCHITECTURAL DESIGN RESEARCH 2

综合设计
COMPREHENSIVE DESIGN

周 凌
ZHOU Ling

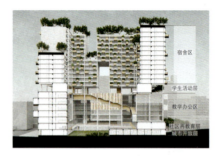

题目：都市再生

研究内容

  基地一：以南京河西奥体地块为例，针对前期城市化过程中遗留下来的不足与部分功能性过时等问题，通过对经济、社会、文化、环境、交通、治理等的分析、研究，提出具有创造性的解决方案，达到地区都市再生与再城市化的目的。

  基地二：以南京大学鼓楼校区周边地块与街道为例，针对人群、商业、场所要素，进行重构再生，以适应新的城市发展需求。

Title: Urban Regeneration

Research content

  Base 1: Taking Nanjing Hexi Olympic Sports Center plot as an example, targeting the problems of deficiencies and partial functional obsolescence left over from the pre-urbanization process, the study proposes creative solutions through analyses and researches on economic, social, cultural, environmental, traffic and governance etc., so as to achieve the purpose of regional urban regeneration and re-urbanization.

  Base 2: Taking the plots and streets surrounding the Gulou Campus of Nanjing University as an example, this study focuses on the reconstruction and regeneration of population, commerce, and place elements to adapt to the new needs of urban development.

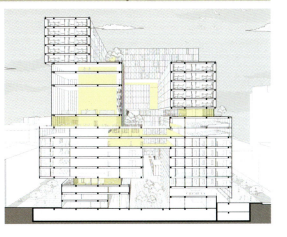

1

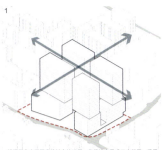

根据概念中不同学院结合的选择，将场地布置成四个组团，采用风车式布局，并通过垂直分层进行功能分区。宿舍区位于上端，教学公共区域在底端。建筑的轮廓参考周围道路和建筑的边界。

2

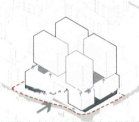

将场地朝向南部的空间打开，建筑间内收缩，形成社区广场，并在教学区域与宿舍区中间增加学生活动空间和公共休憩平台。

3

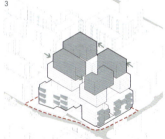

为了满足学生宿舍的采光条件，将宿舍上半部分的体块进行移位，并在教学区域的部增添休息空间，减少阳光直射的同时也增加与周围行人的交互交流。

4

完善建筑的细节部分，并在宿舍外墙增添绿植，使建筑更加绿色生态。

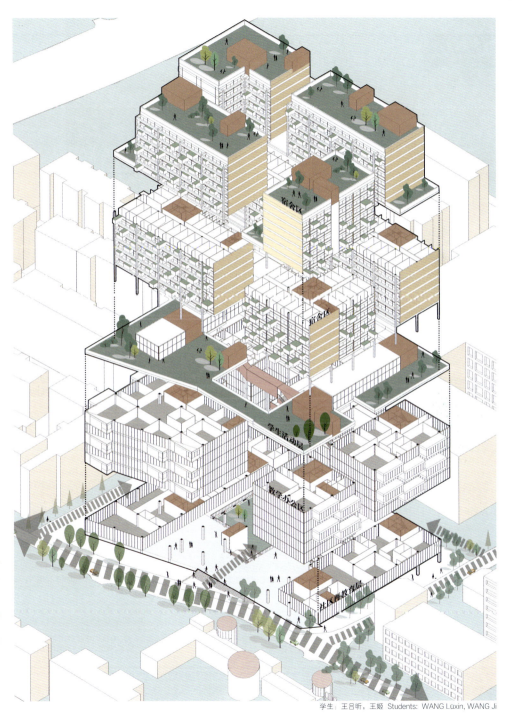

学生：王吕昕，王姬 Students: WANG Lǚxin, WANG Ji

学生：黄佳怡，孙穆群 Students: HUANG Jiayi, SUN Muqun

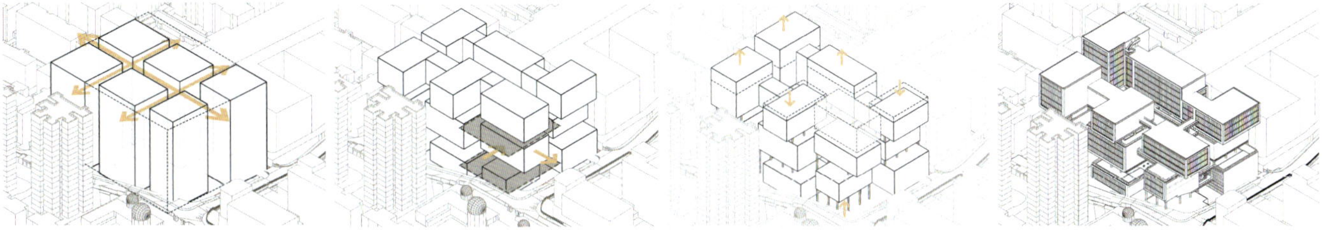

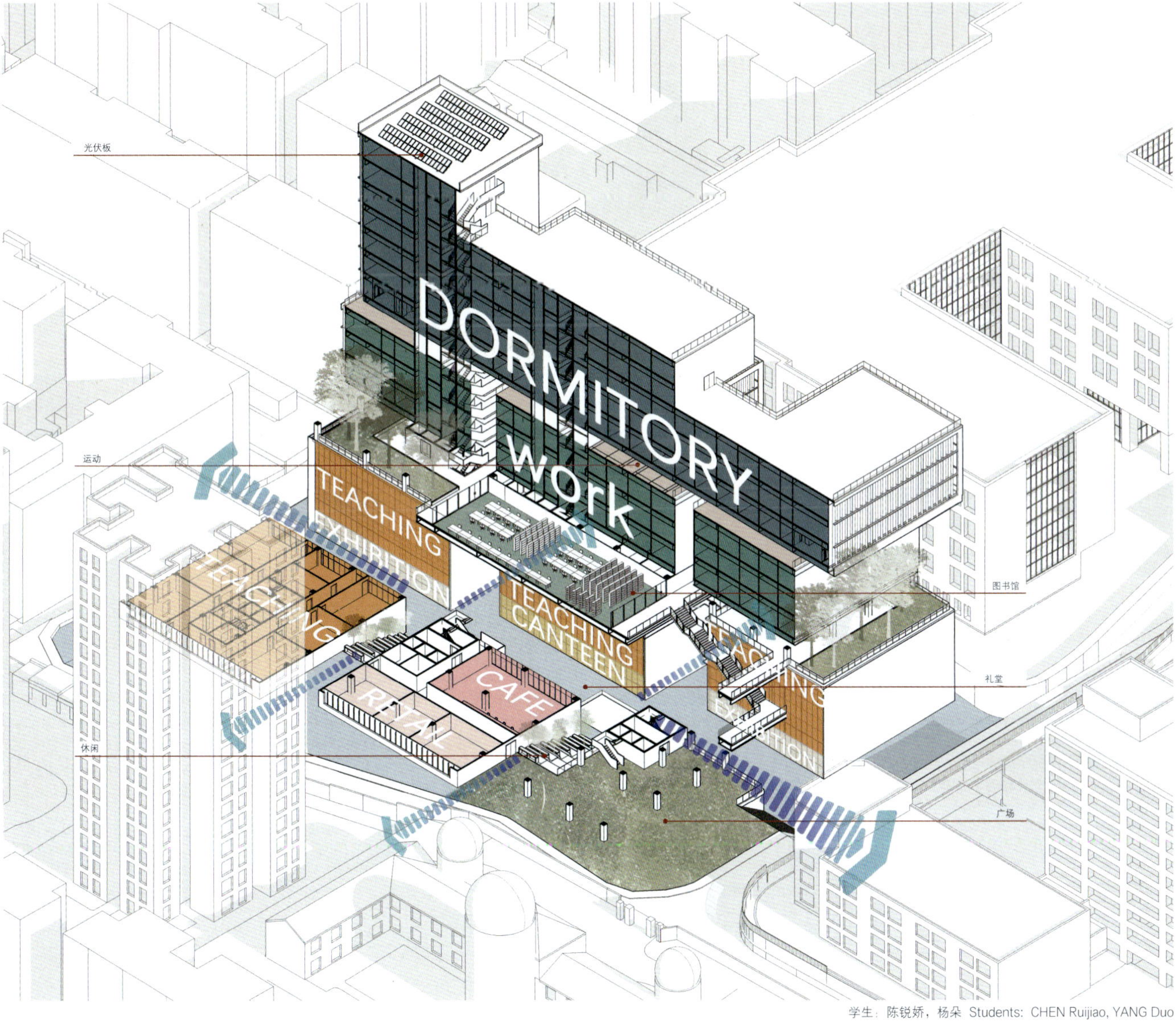

光伏板

运动

休闲

图书馆

礼堂

广场

学生：陈锐娇，杨朵 Students: CHEN Ruijiao, YANG Duo

建筑设计研究（二）ARCHITECTURAL DESIGN RESEARCH 2

综合设计
COMPREHENSIVE DESIGN
金 鑫
JIN Xin

研究课题

城市工业建筑空间再生研究——南京国家领军人才创业园28号楼（原第二机床厂大厂房）改造设计

研究内容

从社会发展的规律性来看，由于工业文明社会向后工业文明社会的转型与过渡，部分工业建筑必然面临停产、搬迁、转移、废弃等境况。而大规模的拆除、重建，必然产生极大的资源浪费、污染排放等问题。因此，以空间再生为基本特征的工业建筑再利用，成为建筑师的新任务。本次设计研究将聚焦工业建筑空间的再利用问题，根据工业建筑空间的特点，结合社会需求的新内容和空间要求，努力探索工业建筑空间再生的规律。

1）尺度转换

工业建筑、构筑物和场地等通常具有远超人体尺度的巨大体量，并容纳大量复杂的机器和设备的运作。这样的物质环境，具有机器化、非人性的尺度和空间，难以与常人的生活、工作等活动相关联。以空间的再生作为设计研究的核心，意味着将工业建筑的巨大空间转向民用、公共的空间。

而工业建筑巨大的空间尺度和坚固的结构体系，提供了重新组织交通流线的可能。这同样需要做相应的空间尺度转换研究，以汽车和车行的尺度作为基本单元来研究工业建筑的空间适应性。

2）程序重置

在原为满足生产工艺流程要求而设置的工业建筑空间及其组织关系中，重新置入符合城市生活需求的新的程序与功能，合理安排新的活动内容。

3）结构重组

在工业建筑的改造中，为了满足空间再利用的需求，可对工业建筑既有结构体系进行改变或重组。新置入的结构体与既有工业建筑结构体系可能形成多种空间位置和受力关系。

课程信息

指导教师：金鑫
合作指导：杨侃
学生人数：21人，三人一组
（结合课程内容进行课堂大组讨论、小组讨论和讲课。）
作业要求：自拟任务书、模型、A0图纸、汇报PPT等
系列讲座拟请嘉宾：
周苏宁（米思建筑）
王子耕（Pills工作室）
马岛（来建筑）
罗宇杰（罗宇杰工作室）

Research topic

Research on Urban Industrial Building Space Regeneration—Renovation Design of Building 28 of Nanjing National Leading Talents Pioneering Park (the formerly large workshop of the Second Machine Tool Factory)

Research content

From the perspective of the law of social development, due to the

transformation and transition from an industrial civilized society to a post-industrial civilized society, some industrial buildings will inevitably face situations such as suspension of production, relocation, transfer, and abandonment. However, large-scale demolition and reconstruction will inevitably lead to huge waste of resources, pollution discharge and other problems. Therefore, the reuse of industrial buildings characterized by space regeneration has become a new task for architects. This design research will focus on the reuse of industrial building space. According to the characteristics of industrial building space, combined with the new content and space requirements of social needs, we will strive to explore the law of industrial building space regeneration.

1) Scale conversion

Industrial buildings, structures, and sites usually have huge volumes that far exceed the scale of the human body, and accommodate the operation of a large number of complex machines and equipment. Such a material environment has a mechanized and inhuman scale and space, and is difficult to relate to ordinary people's life, work and other activities. Taking the regeneration of space as the core of design research means turning the huge space of industrial buildings into civil and public spaces.

The huge spatial scale and solid structural system of industrial buildings provide the possibility to reorganize traffic flow. This also requires corresponding research on spatial scale conversion, using the scale of automobiles and vehicular traffic as the basic unit to study the spatial adaptability of industrial buildings.

2) Program reset

In the industrial building space and its organizational relationship that are originally set up to meet the requirements of the production process, new procedures and functions that meet the needs of urban life are re-installed, and new activities are reasonably arranged.

3) Restructuring

In the transformation of industrial buildings, in order to meet the needs of space reuse, the existing structural system of industrial buildings can be changed or reorganized. The newly placed structure may form a variety of spatial positions and mechanical relationships with the existing industrial building structure system.

Course information

Instructor: Jin Xin

Cooperative mentor: Yang Kan

Number of students: 21 people, in groups of three

(Classroom group discussions, small group discussions and lectures are conducted in conjunction with course content.)

Job requirements: self-created briefs, model, A0 drawing, report PPT, etc.

Invited guests for the lecture series:

Zhou Suning (Misi Architecture)

Wang Zigeng (Pills Studio)

Maldives (Atelier LAI)

Luo Yujie (Luo Yujie Studio)

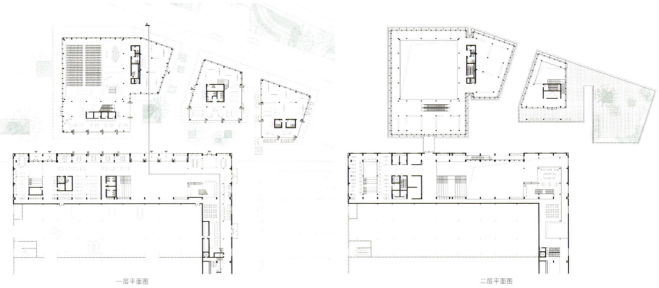

一层平面图　　　　　　　　　　　　　　　二层平面图

剖面图

学生：陈晨，陈雯，沈逸哲 Students: CHEN Chen, CHEN Wen, SHEN Yizhe

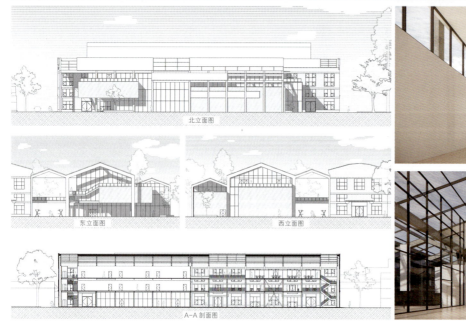

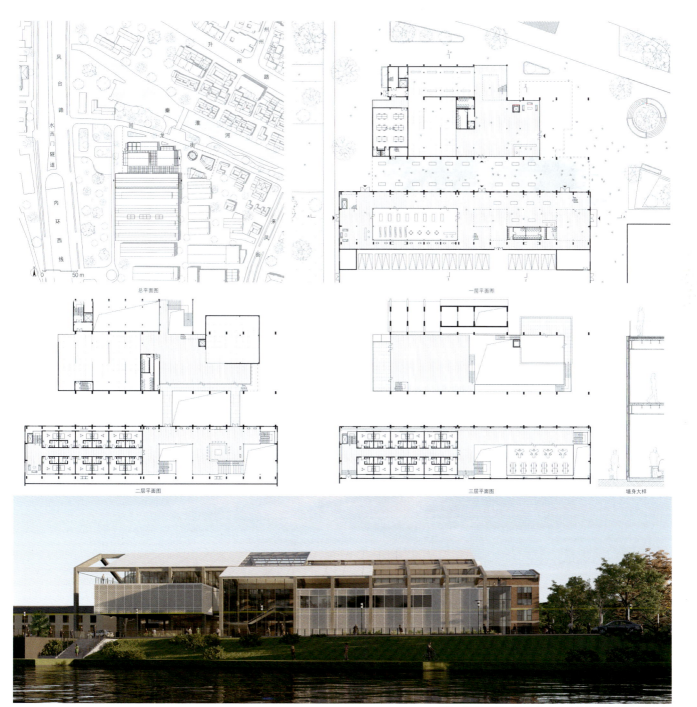

学生：林济武，田靖，喻姝凡 Students: LIN Jiwu, TIAN Jing, YU Shufan

建筑设计研究【二】 ARCHITECTURAL DESIGN RESEARCH 2

综合设计
COMPREHENSIVE DESIGN
程向阳
CHENG Xiangyang

研究课题
　　被置换后老城研究所大院空间再企划

场地概况
　　中国科学院南京土壤研究所位于南京市玄武区北京东路71号，成立于1953年，其前身为1930年创立的中央地质调查所土壤研究室。中国科学院南京地理与湖泊研究所位于南京市玄武区北京东路73号，其前身是1940年8月成立的中国地理研究所，1988年1月改名为中国科学院南京地理与湖泊研究所并沿用至今，是中国唯一以湖泊—流域系统为主要研究对象的地理研究所。两个研究所共用一个大院空间，没有明确的物理边界分割，总占地面积为34 760 m²。

基本问题
　　1）研究所渊源的存续性问题：自中国科学院南京各研究所成立以来，经过数代科研人员的努力，已经有了丰厚的科研成果和文化积淀，具有代表性的比如土壤地力保育与土壤污染修复、太湖蓝藻持续治理及生态修复等科研成果。这次搬迁为办公区整体搬迁，周边家属楼也在房改后进行大量市场化置换，与研究所相关的一切可能会在未来新空间生产过程中消失。
　　2）高需求周边环境与无发展空间的矛盾：研究所的原有地址在老城中紧邻风景区胜地、政府、名校等优质区位和快速交通干道，地铁线路便利，拥有良好的地理位置。周边为密集的居民小区或单位，还没有进入拆迁生命周期，因此该区域处于公共空间使用和配置需求但无空地空间的状态，本次被置换后研究所空间的机遇未知。
　　3）内部建筑空间局限，外部环境识别性弱：研究所的办公区域建筑多为二十世纪五六十年代以后逐步建设的普通办公楼，面临内部空间如何改良、如何界定拆保留的问题。而室外开敞空间少，道路和部分绿化基本被停车占用。研究所被家属区包围，仅余大门与北京东路有空间连接，对外识别性不佳。

规划条件
　　该地段处于南京北京东路与九华山公园之间，是南京老城历史文化、自然山水积淀最为丰富，以及南京城市特色最明显的地区，也是南京紫金山玄武湖中心公园的重要拓展区。对该地段的规划设计不仅要了解总体规划发展目标及相关规划，并关注南京城市特色，还要研判紫金山玄武中心公园以及南京文学之都等计划可能对基地未来图景的影响。该地区规划限高24 m，其他指标可以根据设计合理确定，并符合南京相关规划法规要求（如日照、停车配建等）。

设计任务
　　1）基于城市尺度分析，研究如何对大院搬迁后的空间进行再定义，让后续空间的策划与运营适应新的社会环境和社会需求，并建立可持续的发展模式。
　　2）以既存空间作为研究主线，分析空间在都市中潜在的特质，进行系列的空间实验。
　　3）以空间作为设计实践的对象，架构适配新的内容，激发空间的创造性和价值。

Research topic
　　Spatial Reprogramming of the Displaced Old City Research Institute Compound

Site overview
　　Nanjing Institute of Soil Science, Chinese Academy of Sciences, is located at No. 71 Beijing East Road, Xuanwu District, Nanjing City. It was founded in 1953 and originated from the Soil Research Office of the Central Geological Survey Institute, which was founded in 1930. Nanjing Institute of Geography and Limnology, Chinese Academy of Sciences, is located at No. 73 Beijing East Road, Xuanwu District, Nanjing City. Its predecessor was the Institute of Geography of China, established in August 1940, and it was renamed Nanjing Institute of Geography and Limnology, Chinese Academy of Sciences in January 1988, which has been in use ever since. It is the only geographical research institute in

China with the lake-basin system as the main object of research. The above two institutes share the same compound space without a clear physical boundary division, covering a total area of 34,760 m².

Basic issues

1) Continuity of the origin of the institute: Since the establishment of the various research institutes of the Chinese Academy of Sciences, through the efforts of several generations of scientific researchers, the institute has accumulated abundant scientific research results and cultural deposits, with representative scientific research results such as soil fertility conservation and soil pollution remediation, sustainable management of cyanobacteria in Lake Taihu and ecological restoration, etc. The relocation of this time is an overall relocation of the office area, and the surrounding residential buildings have also been heavily marketed for after the housing reform, so everything related to the research institutes may disappear in the future production process of new spaces.

2) Contradiction between the high demand for the surrounding environment and the lack of space for development: The research institutes' original location in the old city adjacent to scenic resorts, government, famous schools and other high-quality locations, as well as fast traffic arteries, with convenient underground lines, and a good location. Surrounded by dense residential neighborhoods or units, the area has not entered the demolition life cycle, thus it is in a state of having public space usage and configuration needs but without available land. The opportunities for the research institutes' space after this replacement are unknown.

3) Limitations of internal architectural space and weak identification of the external environment: The institutes' office area buildings are mostly ordinary office buildings constructed gradually in the 1950s and 1960s and later, facing the problems of how to improve the internal space and how to define the demolition and preservation. The outdoor open space is less, and the roads and part of the greenery are basically occupied by parking. The research institutes are surrounded by residential areas, and only the main gate is connected to Beijing East Road, which is not well recognized externally.

Planning conditions

This lot is located in the area between Nanjing Beijing East Road and Jiuhuashan Park, which is the area with the richest accumulation of history, culture and natural landscape in the old city of Nanjing as well as the most distinctive area of Nanjing's urban characteristics. It is also the important expansion area for Zijinshan Xuanwu Lake Centre Park. The planning and design of this area not only need to understand the overall planning and development objectives and other relevant plans, and focus on the Nanjing's urban characteristics, but also need to analyze Zijinshan Xuanwu Lake Centre Park and Nanjing Literature Capital and other plans on the future vision of the base. The height limit of this area is 24 m, other indicators can be reasonably determined according to the design, and comply with the requirements of the relevant planning regulations in Nanjing (e.g., sunlight, parking allocation, etc.).

Design tasks

1) Based on urban scale analysis, research on how to redefine the space after the relocation of the compound, let the planning and operation of the subsequent space adapt to the new social environment and social demand, and establish a sustainable development mode.

2) Taking the existing space as the main line of research, analyse the potential qualities of space in the city, and conduct a series of spatial experiments.

3) Taking space as the object of design practice, structure and accommodate new content, stimulate the creativity and value of the space.

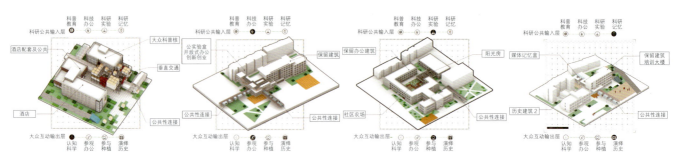

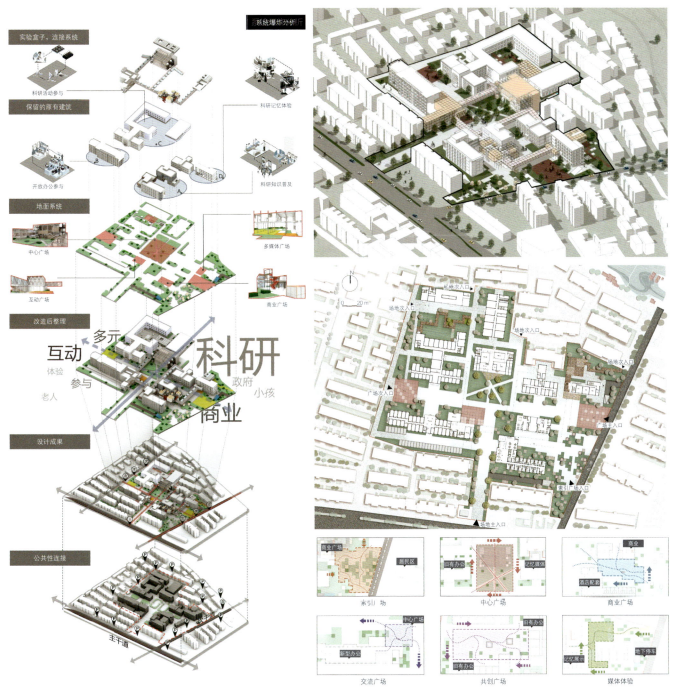

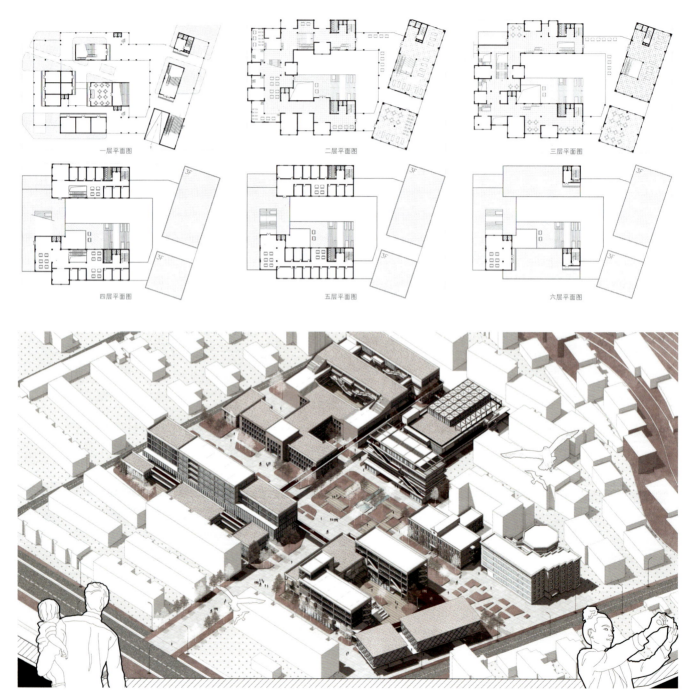

学生：陈哲，贺子琦，奚钰竹，张桂煜　Students: CHEN Zhe, HE Ziqi, XI Yuzhu, ZHANG Guiyu

学生：虞伟炜，王冠一，郭士博，赵潇艺 Students: YU Weiwei, WANG Guanyi, GUO Shibo, ZHAO Xiaoyi

建筑设计研究（二）ARCHITECTURAL DESIGN RESEARCH 2
城市设计
URBAN DESIGN

华晓宁
HUA Xiaoning

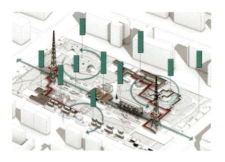

研究课题
　　信息·机器·辖域——信息基础设施的空间政治学（"国民政府中央广播电台"旧址再生）

课程议题
　　基础设施日益成为当代城市研究与实践的重要主题。以往基础设施仅仅被视作市政工程的专业领域，遵循工具理性，被传统建筑学忽视多年。然而基础设施内在具有的独特空间潜力，能够成为重塑空间状态和空间关系的触媒。
　　作为传送、分发信息的机器，信息基础设施的"源""流""域"都具有天然的空间性，赋予空间不同的属性特征。本课题关注城市中的信息基础设施，解读和反思其引发的空间分异，并在当代语境中对其重新定义，创造新的空间关联，从而推动对城市空间的重构。

对象与场址
　　"国民政府中央广播电台"旧址坐落于南京市鼓楼区江东北街 33 号，始建于 1931 年，1932 年 5 月竣工并正式开始播音，直到 2014 年停止使用，期间不断增建改建。电台旧址南北长 207 m，东西宽 302 m，占地面积约 65 224.6 m²，计 97.79 亩。场地内现有文物建筑共 11 处（含两座广播塔），另有多处 1949 年后加建的建筑留存。

设计要求
　　对"国民政府中央广播电台"旧址及其周边地段进行深入调研，探寻该基础设施在城市发展中的潜力，构想该场址的未来愿景，提出场址的再生策略，并选择重点建筑进行改造设计。

Research topic
　　Information · Machine · Territory—The Spatial Politics of Information Infrastructure (Regeneration of the Former Site of the "Nationalist Government's Central Broadcasting Station")

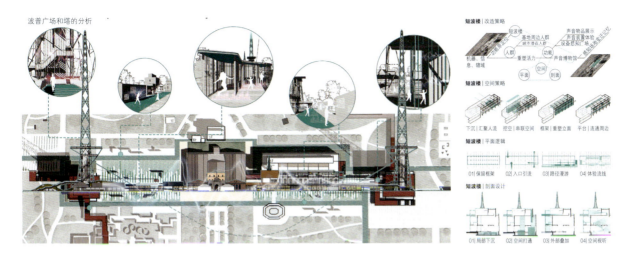

Course topic

Infrastructure has increasingly become an important topic in contemporary urban research and practice. In the past, infrastructure was merely seen as a professional field of municipal engineering, following instrumental rationality, and neglected for many years by traditional architecture. However, infrastructure inherently possesses unique spatial potential, which can serve as a catalyst for reshaping spatial conditions and spatial relationships.

As a machine for transmitting and distributing information, the "source," "flow," and "domain" of information infrastructure all have inherent spatiality, endowing space with different attribute characteristics. This study focuses on information infrastructure in cities, interpreting and reflecting on the spatial differentiation it triggers, and redefining it in the contemporary context, creating new spatial relationships to drive the reconstruction of urban space.

Objects and sites

The former site of the "Nationalist Government's Central Broadcasting Station" in Nanjing is located at No. 33 Jiangdongmen North Street, Gulou District, Nanjing City. It was built in 1931 and completed in May 1932, officially beginning broadcasting. It was in use until 2014, during which time there were continuous additions and renovations. The site of the radio station is 207 meters long from north to south and 230 m wide from east to west, covering an area of approximately 65,224.6 m^2, or 97.79 acres. There are a total of 11 existing historical buildings on the site (including two broadcasting towers), as well as several buildings added after 1949.

Design requirements

Conduct in-depth research on the former site of the "Nationalist Government's Central Broadcasting Station" and its surrounding areas, explore the potential of this infrastructure in urban development, envision the future vision of the site, propose strategies for its revitalization, and select key buildings for renovation and design.

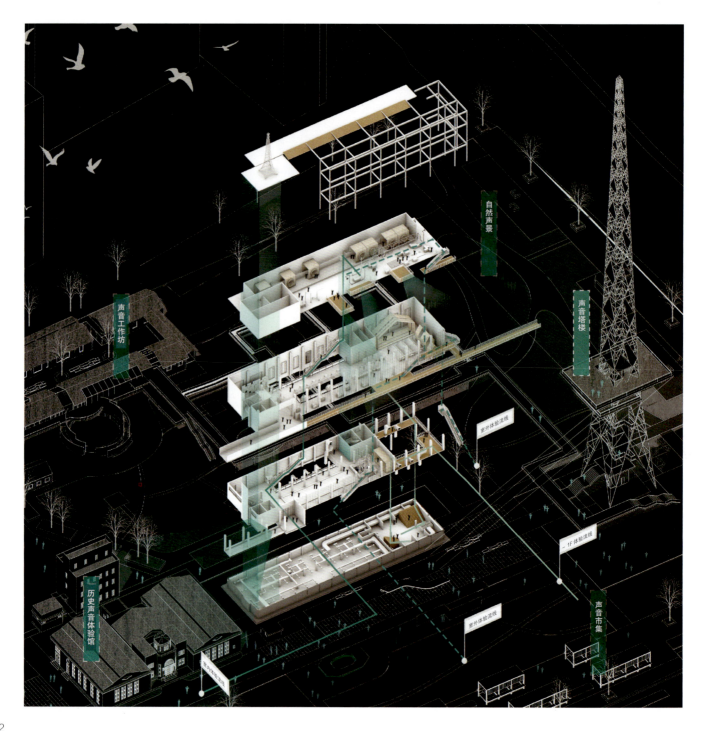

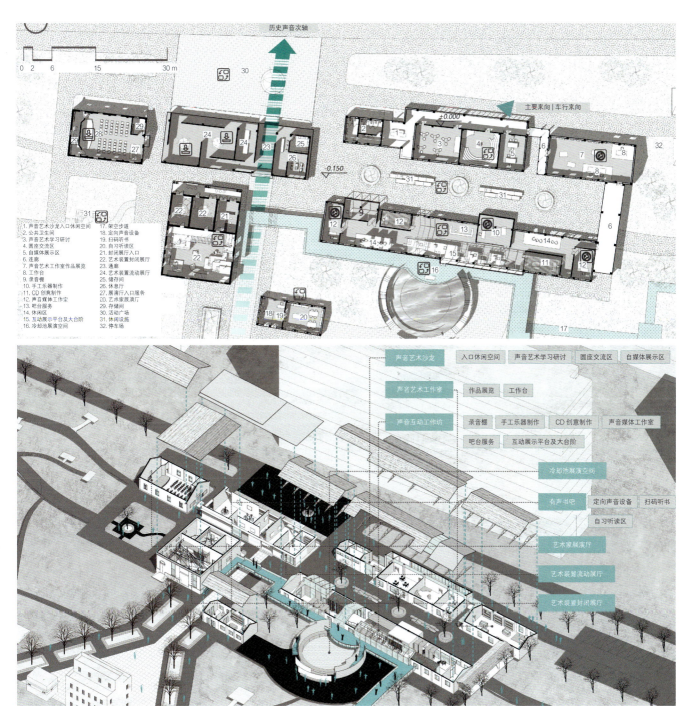

学生：李雨茜，洪倩倩，汪晶，王姬 Students: LI Yuxi, HONG Qianqian, WANG Jing, WANG Ji

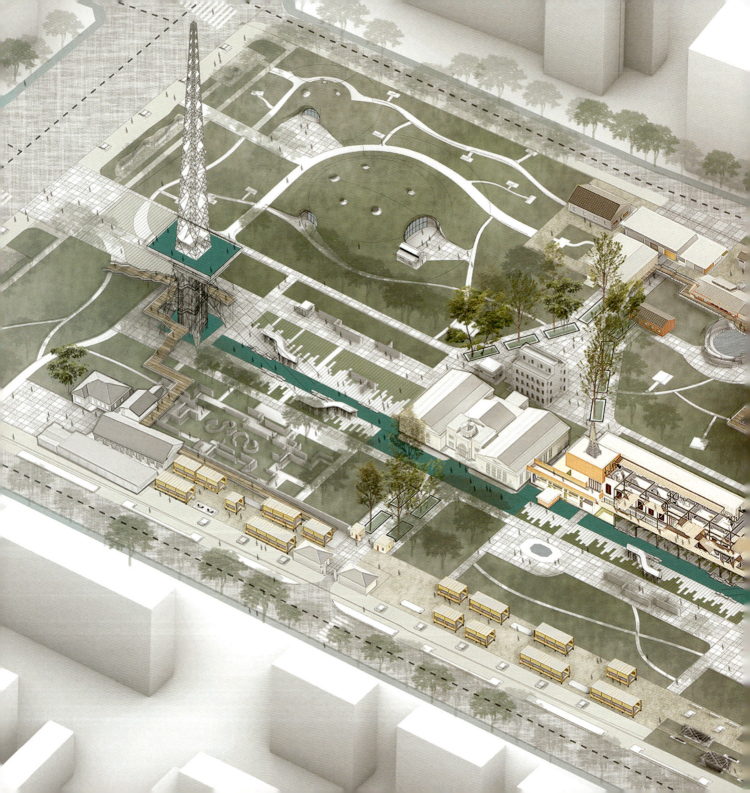

建筑设计研究（二）| ARCHITECTURAL DESIGN RESEARCH 2

城市设计
URBAN DESIGN
胡友培
HU Youpei

研究课题

都市区重构——一个关于开放城市结构的设计试验

背景与议题

当下中国城市的发展进入都市区阶段，在稠密的传统中心城区之外是尺度巨大的都市区域。在城市存量发展与更新背景下，这很可能是未来城市化深耕的主战场。

相比于成熟的传统城区，都市区域的特征之一是快速而野蛮的生长导致发展形态碎片化，各种异质的要素毫无征兆地碰撞在一起，产生出混乱但生机勃勃的都市景观，即所谓的城乡接合部。针对该特征，应考虑如何在碎片化的现状中重构出某种都市空间秩序或形态结构，为松散粗放的都市环境注入某种都市性。

都市区域的特征之二是变动不定以及不确定的未来。这导致无法运用蓝图式规划思维加以应对，任何构图式的总体规划都注定失效。针对该特征，应考虑是否存在一种动态的、开放的城市结构与形态，以容纳都市区不确定的未来与动态。

都市区域的特征之三是粗放圈地开发已经基本成型，全盘推翻毫无可能，只能以城市更新的方式修修补补，这无疑为秩序的重构带来极大的难度。针对该特征，应考虑如何以最低的成本，在保持全局视野的基础上，选取战略性点位进行更新与重构。

设计命题与任务

本次课程的命题与任务是以设计试验的方式，在混乱而粗放的都市区域中，采用战略点位更新的策略，探索一种开放的城市结构形态，容纳未来的不确定和动态，重构混乱无序的都市区景观。

场地与范围

研究范围拟定为南京桥北区域，南京大学金陵学院周边约 3 km×3 km 的范围。该区域具有都市区外围地段典型的城乡接合部景观。场地边界、规模、范围通过研究自行设定。

课程限定与要求

1）建筑的介入

开放的结构形态，不是城市规划中使用的抽象的宏观拓扑结构，而是建筑学尺度的、可被人认知的空间结构。要求以建筑的方式介入都市区的空间生产与城市更新中，探索在建筑尺度上，以最低的空间生产成本进行城市更新，形成开放结构的可行路径。

2）设计方法论

设计试验需要相对严谨的方法论。对于未知开放的议题，要求以一种科学实证的精神，而不仅仅是形式主义的游戏，来推演开放结构的可能形态，并通过设定的方法论与操作流程，以抵抗经验和美学的过早干预，而抵达未知和新颖的成果，并对结果的开放性做出阐明。

Research topic

Metropolitan Area Restructuring—A Design Experiment on Open Urban Structure

Background and topic

The current development of Chinese cities has entered the metropolitan area stage. Beyond the dense traditional central city areas, there are vast metropolitan areas. In the context of urban stock development and renewal, this is likely to be

the main battlefield for future urbanization.

Compared to mature traditional urban areas, the first characteristic of metropolitan areas is rapid and brutal growth, leading to fragmented development forms, various heterogeneous elements, and sudden collisions that create chaotic but vibrant urban landscapes, known as urban periphery. Aiming at this characteristic, consideration should be taken to how to reconstruct some form of urban spatial order or morphological structure in the fragmented current situation, injecting a sense of urbanity into the loose and extensive urban environment.

The second characteristic of metropolitan areas is their variability and uncertain future. This makes it impossible to respond with blueprint-style planning thinking, as any master planning approach is destined to fail. Aiming at this characteristic, it should be considered whether there exists a dynamic and open urban structure and morphology that can accommodate the uncertain future and dynamics of metropolitan areas.

The third characteristic of metropolitan areas is that extensive land development has already taken shape and cannot be completely overturned. It can only be repaired and patched up through urban renewal, which undoubtedly brings great difficulty to the reconstruction of order. Aiming at this characteristic, consideration should be taken to how to select strategic points for renewal and reconstruction with the lowest cost while maintaining a comprehensive vision.

Proposition and task design

The proposition and task of this course is to explore an open urban structure form that can accommodate future uncertainties and dynamics, and reconstruct the chaotic and disorderly urban area landscape through the use of design experience and the strategy of updating strategic point in chaotic and sprawling urban areas.

Site and scope

The research scope is proposed to be the Northen Bridge area of Nanjing, with a surrounding area of approximately 3 km × 3 km around Jinling College of Nanjing University. This area features a typical urban fringe landscape. The site boundaries, scale, and scope will be self-determined through research.

Course restrictions and requirements

1) Architectural intervention

An open structural form is not the abstract macro topological structure commonly used in urban planning, but a spatial structure at the scale of architecture that is recognizable by people. It is required to intervene in urban space production and city renewal in a way that is architectural, explore to conduct city renewal in the most cost-effective way at the architectural scale, and form a feasible path of open structure.

2) Design methodology

Designing experiments require a relatively rigorous methodological framework. For unknown open topics, it is required to deduce possible forms of open structure with a scientific and empirical spirit rather than just playing with formalism, resist premature intervention of experience and aesthetics through the set methodology and operational processes to reach unknown and novel outcomes, and clarify the openness of the results.

学生：初晓畅，王琪琪，孙鹤，孙强，胡珊珊 Students: CHU Xiaochang, WANG Qiqi, SUN He, SUN Qiang, HU Shanshan

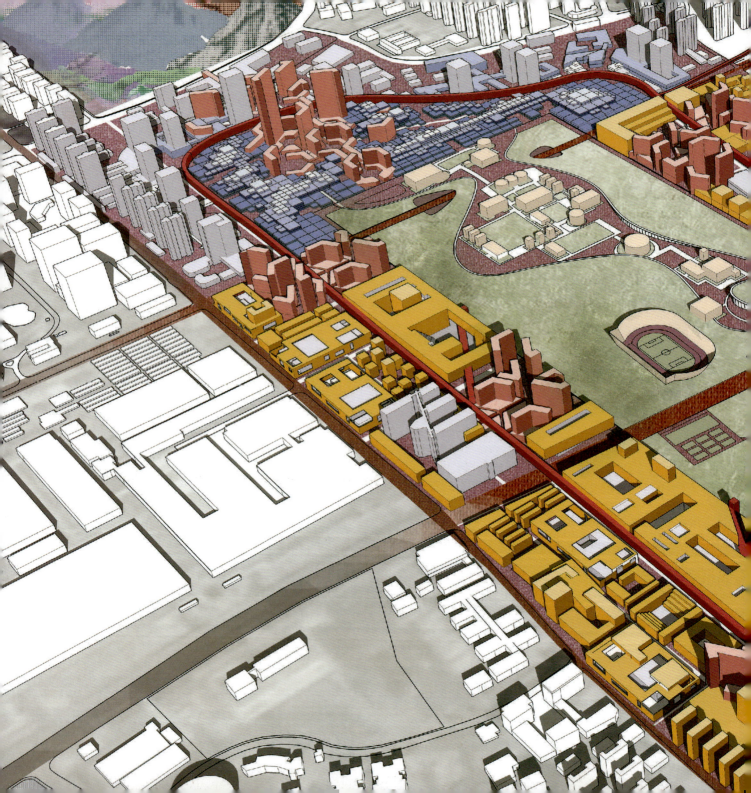

研究生国际教学工作坊 POSTGRADUATE INTERNATIONAL DESIGN STUDIO

回收边缘——作为城市设备的校园
RECYCLING THE EDGES—THE CAMPUS AS AN URBAN DEVICE
保拉·佩莱格里尼
Paola PELLEGRINI

背景与议题

联合国人居署推荐采取回收、再利用和减少行动以实现可持续发展和应对气候变化。这些行动应该影响城市的规划和设计方式，因为城市既是问题的一部分，也是解决方案的一部分。

全世界的设计师都在重新考虑已经建造的东西，以提高其性能和质量，同时节省金钱、能源、土地和自然资源。这种方法通常被称为"再生"，即将建成环境转变为更好或更有价值的状态。可持续再生——尽可能减少对环境的负面影响——是一个挑战，它需要仔细观察和创造性地寻找新的机会和设计方式，以适应现存城市。工作坊的练习以大学校园及其周边的城市空间为目标。学生必须观察校园边缘的空间——校园内外、现有建筑和开放空间。学生必须设想如何改变边缘位置，增加新的用途和建筑结构，使空间更加高效。挑战在于理解校园如何成为城市设备，即如何提供空间和结构，以供校园学生和职员以及市民使用，特别是周边区域的居民。

校园的边缘也很重要，因为它们给予了大学特色和身份，特别是入口。目标是重新思考和设计，以增加和多样性场所的供应，并提高它们的视觉冲击力和质量。

课程限定与要求

学生必须确定一个小空间，了解为什么它未被使用，定义新用途和用户的项目（将用于什么？谁将使用这个空间？），起草设计以提高其性能。城市设计尤其注重开放的公共空间，因此设计中也必须包含一个开放区域。

Background and topic

The UN Habitat recommends actions of recycling, reusing, reducing for a sustainable development and for tackling climate change. These actions should impact the way the city is planned and designed, because cities are both part of the problem and part of the solution.

All over the world designers are reconsidering what is already built to improve its performance and quality while saving money, energy, land and natural resources. This approach is usually called "regeneration", when the built environment is transformed to a better or more worthy state. Sustainable regeneration—one that reduces as much as possible the negative impacts on the environment—is a challenge that requires careful observation and creativity for finding new opportunities and design approaches in the existing city. The exercise of the workshop targets the university campus and its edges with the surrounding urban spaces. The students must observe the spaces along the edges of the campus—both inside and outside the campus, both the existing buildings and open spaces. Students must imagine where and how the edges can be transformed and new uses and building structures can be added to make the space more efficient. The challenge is to understand how the campus can become an urban device, meaning how it can offer spaces and structures for uses and activities that can benefit both the campus' students and staff and the citizens, in particular the residents of the surrounding areas.

The edges of the campus are important because they play a role in giving a character and an identity to the university, gateways in particular. The goal is to rethink and redesign to increase and diversify the offer of places and improve their visual impact and quality.

Course restrictions and requirements

Students must identify a small space, understand why it is under-used, define a program of new uses and users (what is it going to be used for? Who will use the space?), draft a design to increase its performance. Urban design pays particular attention to open public space so there must be also an open area in the design.

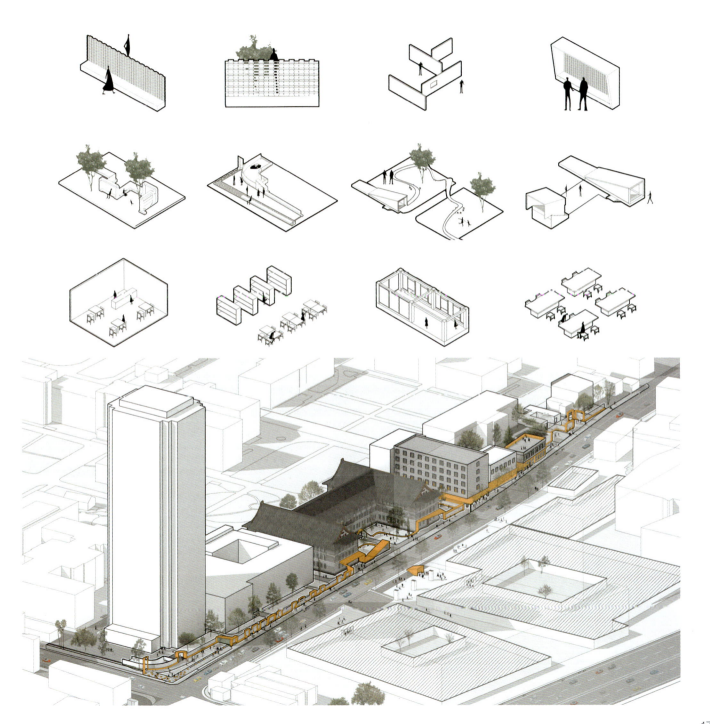

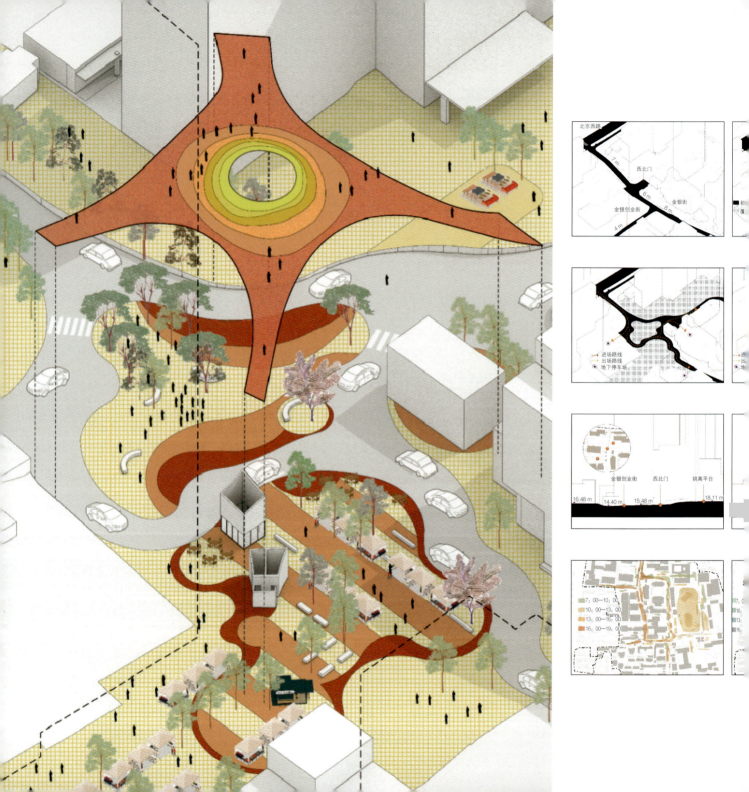

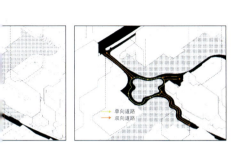
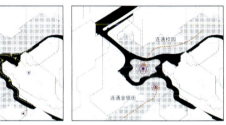
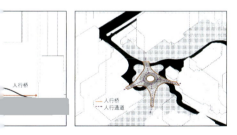

学生：帕·普纳克莱塞，马子昂，赵文，杨茸佳，骆婧雯 Students: Pa PONNAKRAINGSEY, MA Zi'ang, ZHAO Wen, YANG Rongjia, LUO Jingwen

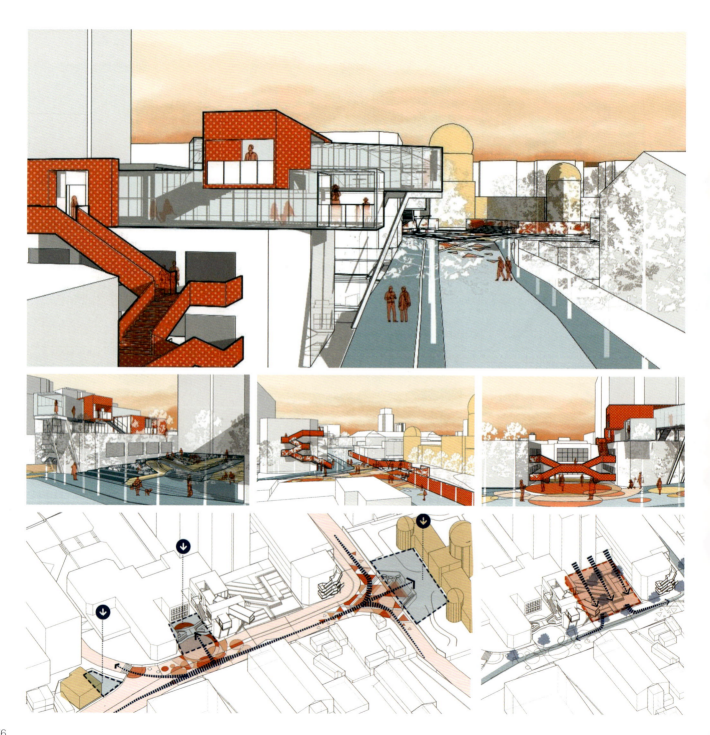

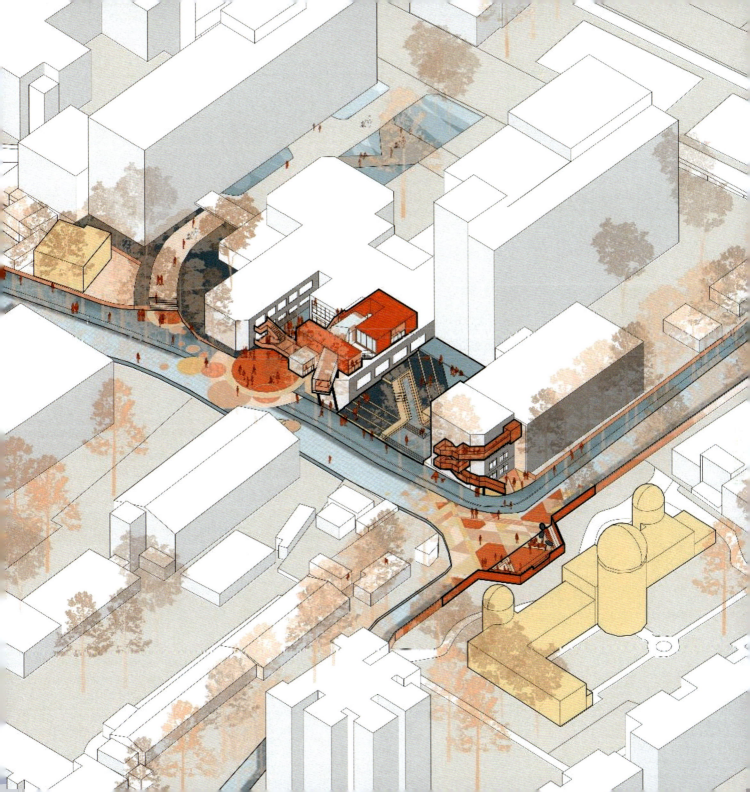

研究生国际教学工作坊 POSTGRADUATE INTERNATIONAL DESIGN STUDIO

可持续建筑设计
SUSTAINABLE ARCHITECTURAL DESIGN
何塞·阿莫多瓦尔 唐莲
Jose ALMODOVAR TANG Lian

教学目标

本课程介绍学生如何设计可持续发展建筑和使用不同的环境设计工具。通过综合分析和计算机模拟物理环境、建筑类型、朝向、内部配置、围护结构的影响,为学习和探索新的设计概念和策略提供了跨学科和跨文化的环境。环境设计的概念和方法将指导整个设计过程。学生将考虑当地和全球的文化价值观,提出高能源效率和环境质量的创新建议。该课程包括两个部分:系列讲座和设计工作室,强调在做中学。

课程要求

要求学生根据课程计划展示他们的工作材料。最终演示应包含 A0 图板,展示不同设计方案的分析方案、计算机模拟图、比较表、规划图和立面图、最终方案的透视图/轴测图、关于设计过程和概念讨论的团队说明、项目能源和环境性能的数据汇总。

Teaching objectives

This course introduces students to how to design sustainable buildings and use different environmental design tools. It provides a cross-disciplinary and cross-cultural environment for learning and exploring new design concepts and strategies through integrated analysis and computer simulations of effects of physical environment, building typology, orientation, internal configuration, thermal envelope. The concepts and methods of environmental design will guide the whole design process. Students will take into account local and global cultural values to make innovative proposals of high energy efficiency and environmental quality. The course includes two parts: lecture series and design studio that emphasis on learning by doing.

Course requirements

Students are required to present their working material according to the course program. The final presentation should contain A0 panels showing analytical schemes of different design options, computer simulation diagrams, comparative tables, plans and sections, perspective/axonometric of final proposal, team statement about the design process and conceptual discussion, data summary of the energy and environmental performance of the project.

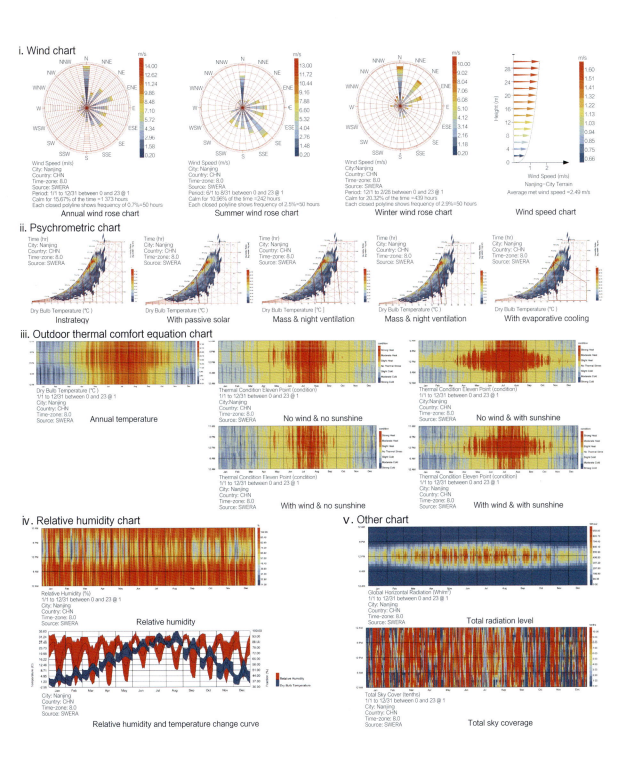

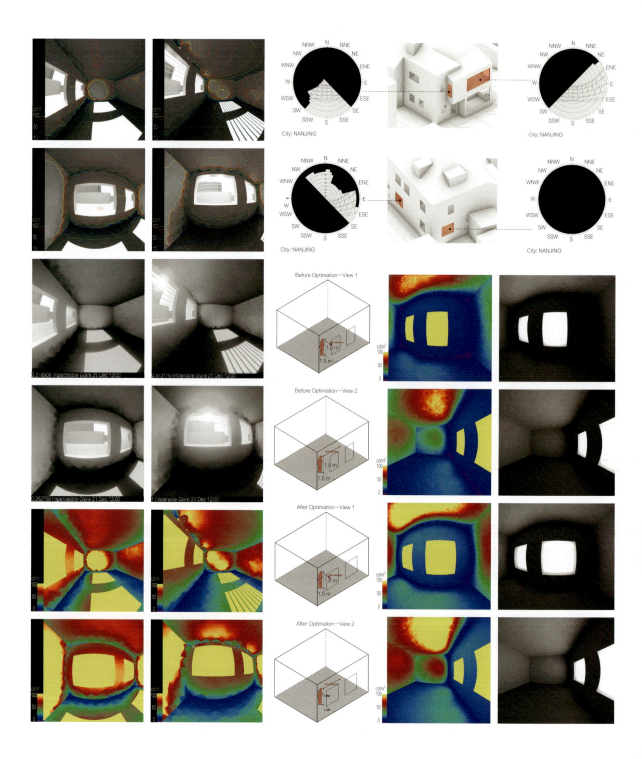

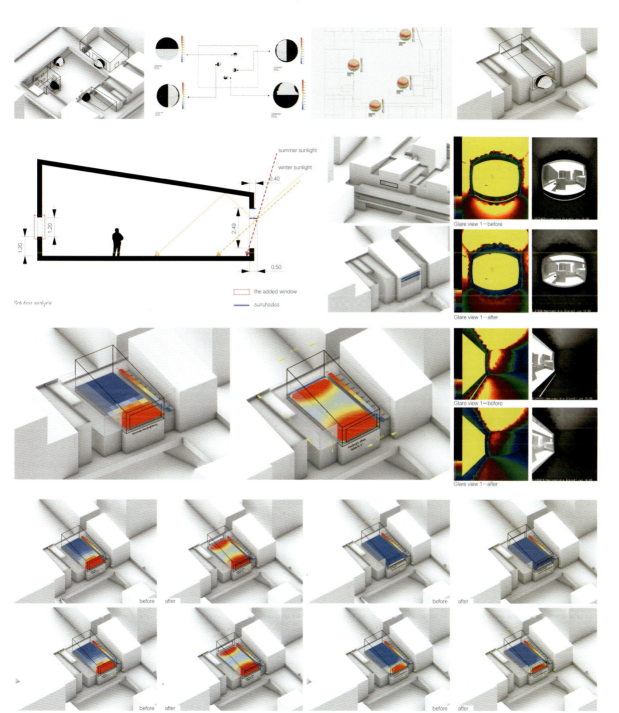

学生：孙强，刘传，刘佳慧，彭幼妹，姜辰达 Students: SUN Qiang, LIU Chuan, LIU Jiahui, PENG Youmei, JIANG Chenda

建筑设计课程
ARCHITECTURAL DESIGN COURSES

本科一年级
设计基础
刘铨 黄春晓 史文娟 梁宇舒
课程类型：必修
学时学分：64学时 / 2学分

Undergraduate Program 1st Year
DESIGN FOUNDATION · LIU Quan, HUANG Chunxiao, SHI Wenjuan, LIANG Yushu
Type: Required Course
Study Period and Credits: 64 hours / 2 credits

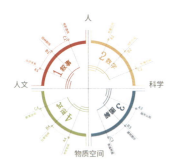

课程内容
第一阶段：知觉、再现与设计
知识点：人与物——材料的知觉特征（不同状态下的色彩、纹理、平整度、透光性等）与物理化学特征（成分、质量、力学性能等）；摄影与图片编辑——构图与主题、光影与色彩；三视图与立面图绘制；排版及其工具——标题、字体、内容主次、参考线。
第二阶段：需求与设计
知识点：人与空间——空间与尺度的概念、行为、动作与一个基本空间单元或空间构件尺寸的关系；平面图、剖面图、轴测图的绘制；线型、线宽、图幅、图纸比例、比例尺、指北针、剖断符号、图名等的规范绘制。
第三阶段：制作与设计
知识点：物与空间——建构的概念；空间的支撑、包裹与施工；实体模型制作——简化的建造；计算机建模工具——虚拟建造；透视图绘制。
第四阶段：环境与设计
知识点：人、物与空间——城市形态要素、城市肌理与城市外部空间的概念；街道系统与交通流线；土地划分与功能分类；总平面图、环境分析图（图底关系、交通流线、功能分区、绿地景观系统）；照片融入表达。

Course content
Phase one: Perception, representation and design
Knowledge points: People and objects—the perceptual characteristics of materials (color, texture, flatness, light transmittance, etc. in different states) and physical and chemical characteristics (composition, quality, mechanical properties, etc.); photography and picture editing—composition and theme, light, shadow and color; three-view and elevation drawing system; typography and its tools—headings, fonts, primary and secondary content, and reference lines.
Phase two: Requirements and design
Knowledge points: People and space—the concept of space and scale, the relationship between behavior, action and the size of a basic spatial unit or spatial component; plan, section, and axonometric drawings; standardized drawing of line type, line width, map size, drawing scale, scale bar, compass, section symbol, drawing title, etc.
Phase three: Production and design
Knowledge points: Objects and space—the concept of construction; space support, wrapping and construction; physical model making—simplified construction; computer modeling tool—virtual construction; perspective drawing.
Phase four: Environment and design
Knowledge points: People, objects and space—the concept of urban form elements, urban texture and urban external space; street system and traffic flow; land division and functional classification; general plan, environmental analysis map (relationship between map and ground, traffic flow, functional zoning, green space landscape system); integration of images in expression.

本科二年级 / 建筑与规划设计（一）（实验班课程）
建筑设计（一）：独立居住空间设计
刘铨 杨舢 唐莲 何仲禹 吴佳维
课程类型：必修
学时学分：64学时 / 4学分

Undergraduate Program 2nd Year
ARCHITECTURAL DESIGN 1: INDEPENDENT LIVING SPACE DESIGN · LIU Quan, YANG Shan, TANG Lian, HE Zhongyu, WU Jiawei
Type: Required Course
Study Period and Credits: 64 hours / 4 credits

课程内容
本次练习的主要任务是综合运用前期案例学习中的知识点——建筑在水平方向上如何利用高度、开洞等操作划分空间，内部空间的功能流线组织及视线关系，墙身、节点、包裹体系、框架结构的构造方式，周围环境对空间、功能、包裹体系的影响等，初步体验一个小型独立居住空间的设计过程。

教学要点
1. 场地与界面：本次设计的场地面积在80—100 m²，场地单面或相邻两面临街，周边为1—2层的传统民居。
2. 功能与空间：本次设计的建筑功能为小型家庭独立式住宅（附设有书房功能）。家庭主要成员包括一对年轻夫妇和1—2位儿童（7岁左右）。新建建筑面积160—200 m²，建筑高度 ≤ 9 m（不设地下空间）。设计者根据设定的家庭成员的职业及兴趣爱好确定空间的功能（职业可以是但不局限于理、工、医、法的技术人员）。
3. 流线组织与出入口设置：考虑建筑内部流线合理性以及建筑出入口与场地周边环境条件的合理衔接。
4. 尺度与感知：建筑中的各功能空间的尺寸需要以人体尺度及人的行为方式作为基本参照，并通过图示表达空间构成要素与人的空间体验之间的关系。

Course content
The main task of this exercise is to comprehensively use the knowledge points in the early case study—how to use height, opening and other operations to divide space in the horizontal direction of the building, functional streamline organization and line of sight relationship of internal space, construction mode of wall body, nodes, wrapping system and frame structure, the influence of the surrounding environment on the space, function and wrapping system, and preliminarily experience the design process of a small independent living space.

Teaching essentials
1. Site and interface: The site of this design covers an area of about 80–100 m², facing the street on one side or two adjacent sides, surrounded by traditional residential buildings of 1–2 floors.
2. Function and space: The building function of this design is a small family independent residence (with study function attached). The main members of the family include a young couple and 1–2 children (about 7 years old). The new building area is 160–200 m² and the building height is ≤ 9 m (no underground space). The designer determines the function of the space according to the set occupation and interests of family members (the occupation can be but not limited to technicians of science, engineering, medicine and law).
3. Streamline organization and entrance and exit setting: Consider the rationality of the internal streamline of the building and the reasonable connection between the entrance and exit of the building and the surrounding environmental conditions of the site.
4. Scale and perception: The size of each functional space in the building needs to take the human body scale and human behaviors as the basic reference, and express the relationship between spatial constituent elements and human spatial experience through diagrams.

本科二年级
建筑设计（二）：校园多功能快递中心设计
冷天 何仲禹 吴佳维 孟宪川
课程类型：必修
学时学分：64学时 / 4学分

Undergraduate Program 2nd Year
ARCHITECTURAL DESIGN 2: CAMPUS MULTIFUNCTIONAL EXPRESS CENTER DESIGN · LENG Tian, HE Zhongyu, WU Jiawei, MENG Xianchuan
Type: Required Course
Study Period and Credits: 64 hours / 4 credits

课程内容

在社会信息化、电商化程度日益提高的背景下，"快递"活跃且丰富地改变了人们的日常生活，成为保障基本生活需求的重要方式。其中，高校内的快递行为富集，快递中心渐渐成为校园后勤服务中不可或缺的一环，与师生日常活动密不可分，成为校园生活区像食堂、公共浴室、超市一样重要的基础公共设施。本次练习的主要任务是在南京大学鼓楼校区南园建设一个校园多功能快递服务中心，要求综合运用建筑设计基础课程的知识点，操作一个小型公共建筑设计项目。

Course content

Against the background of increasing social informatization and e-commerce, "express delivery" has actively and richly changed people's daily lives and become an important way to ensure basic living needs. Among them, express delivery behaviors in colleges and universities are enriched, and express delivery center has gradually become an indispensable part of campus logistics services. It is inseparable from the daily activities of teachers and students, and has become an important basic public facility in campus living areas like canteens, public bathrooms, and supermarkets. The main task of this exercise is to build a campus multi-functional express center in the Nanyuan of Gulou Campus of Nanjing University. It is required to comprehensively use the knowledge points of the basic course of architectural design to operate a small public building design project.

本科二年级 / 本科三年级
建筑设计（三）：专家公寓设计
童滋雨 窦平平 黄华青 钟华颖
课程类型：必修
学时学分：72学时 / 4学分

Undergraduate Program 2nd Year/ 3rd Year
ARCHITECTURAL DESIGN 3: DESIGN OF EXPERT APARTMENTS · TONG Ziyu, DOU Pingping, HUANG Huaqing, ZHONG Huaying
Type: Required Course
Study Period and Credits: 72 hours / 4 credits

课程内容

拟在南京大学鼓楼校区南园宿舍区内新建专家公寓一座，用于国内外专家到访南京大学开展学术交流活动期间的居住。用地位于南园中心喷泉西侧，面积约3 600 m²。地块上原有建筑将被拆除，新建筑总建筑面积不超过3 000 m²。高度不超过3层。

教学目标

从空间单元到系统的设计训练。

从个体到整体，从单元到体系，是建筑空间组织的一种基本和常用方式。本课题首先关注空间单元的生成，并进一步根据内在的使用逻辑和外在的场地条件，将多个单元通过特定方式与秩序组合起来，形成一个兼具合理性、清晰性和丰富性的整体系统。基本单元的重复、韵律、变异等都是常用的操作手法。

Course content

It is proposed to build a new expert apartment in the Nanyuan dormitory area of Gulou Campus of Nanjing University for domestic and foreign experts to live during their visit to Nanjing University for academic exchange activities. The site is located in the west of the fountain in the center of Nanyuan, covering an area of about 3,600 m². The original buildings on the plot will be demolished, and the total construction area of the new buildings will not exceed 3,000 m². The height shall not exceed 3 floors.

Teaching objectives

Design training from space unit to system.

From individual to whole, and from unit to system, it is a basic and common way of architectural space organization. This topic first pays attention to the generation of spatial units, and further combines multiple units with order in a specific way according to the internal use logic and external site conditions to form an overall system with rationality, clarity and richness. Repetition, rhythm and variation of basic units are commonly used

本科二年级 / 本科三年级
建筑设计（四）：世界文学客厅
华晓宁　尹航　梁宇舒
课程类型：必修
学时学分：72 学时 / 4 学分

Undergraduate Program 2nd Year/ 3rd Year
ARCHITECTURAL DESIGN 4: WORLD LITERATURE LIVING ROOM · HUA Xiaoning, DOU Pingping, HUANG Huaqing
Type: Required Course
Study Period and Credits: 72 hours / 4 credits

课程内容

南京古称金陵、白下、建康、建邺……历来是人文荟萃、名家辈出之地，号称"天下文枢"。南京作为六朝古都，亦为中国文学之始。何为文？梁元帝曰："吟咏风谣，流连哀思者，谓之文。"汉魏有文无学，六朝文学《文选》《文心雕龙》《诗品》既是文学评论的开始，也是文学的发端。

2019 年，南京入选联合国"世界文学之都"，开展一系列城市空间计划，包括筹建"世界文学客厅"，作为一座以文学为主题的综合性博物馆。该馆计划位于台城花园与台城城墙之间，解放门西侧，用地面积约 6 500 m²，毗邻古鸡鸣寺、玄武湖、明城墙等历史文化遗迹，构成城市与山林之间的过渡空间。设计应妥善处理建筑与周边城市环境和既有建筑的关系，彰显中国文学的精神特质。

教学目标

本课程主题是"空间"，学习建筑空间组织的技巧和方法，训练对空间的操作与表达。空间问题是建筑学的基本问题。本课题基于文学主题，训练文本、叙事与空间序列的串联，学习空间叙事与空间用途的整体构思，充分考虑人在空间中的行为、空间感受，尝试以空间为手段表达特定的意义和氛围，最终形成一个完整的设计。

Course content

In 2019, Nanjing was selected as a "City of Literature" by the United Nations and planed to carry out a series of urban space plans, including the preparation for the construction of the "World Literature Living Room" as a comprehensive museum with literature as the theme. The museum is planned to be located between Taicheng Garden and Taicheng City Wall, on the west side of Jiefang Gate, covers an area of about 6,500 m², adjacent to ancient Jiming Temple, Xuanwu Lake, Ming City Wall and other historical and cultural relics, forming a transitional space between the city and the mountains. The design should properly deal with the relationship between the building and the surrounding urban environment and existing buildings, and highlight the spiritual characteristics of Chinese literature.

Teaching objectives

The theme of this course is "space", learning the skills and methods of architectural space organization, and training the operation and expression of space. The space problem is the basic problem of architecture. Based on the literary theme, this topic trains the series of the text, narration and spatial sequence, learns the overall idea of spatial narration and spatial use, fully considers people's behaviors and spatial feelings in space, tries to express specific meaning and atmosphere by means of space, and finally forms a complete design.

本科三年级
建筑设计（五）：大学生健身中心改扩建设计
傅筱　钟华颖　孟宪川
课程类型：必修
学时学分：64 学时 / 4 学分

Undergraduate Program 3rd Year
ARCHITECTURAL DESIGN 5: RECONSTRUCTION AND EXPANSION DESIGN OF COLLEGE STUDENT FITNESS CENTER · FU Xiao, ZHONG Huaying, MENG Xianchuan
Type: Required Course
Study Period and Credits: 64 hours / 4 credits

教学目标

本项目拟在南京大学鼓楼校区体育馆基地处改扩建大学生健身中心，以服务于南京大学师生，可适当考虑对周边居民的服务。根据基地条件、功能要求进行建筑和场地设计。

本课题以大学生健身中心为训练载体，在学习中小跨建筑的基本设计原理的基础上，理解建筑形式、空间与基本的结构类型与之间的逻辑关系，初步培养学生整合建筑内外空间、建筑结构、建筑场地，以及兼顾合理的建筑采光通风节能的综合能力。

教学内容

现状基地由一座体育馆和一座游泳馆（吕志和馆）组成，设计需保留体育馆，拆除现状游泳馆并重新设计一座大学生健身中心，其建筑总面积约 4 300 m²（可上下浮动 10%）。建筑高度控制在 24 m 以下，注意场地东西向高差，场地下挖不得超过一层，深度不超过 4.5 m。

Teaching objectives

This project intends to rebuild and expand the college student fitness center at the gymnasium base of the Gulou Campus of Nanjing University to serve the teachers and students of Nanjing University, and give due consideration to the service of surrounding residents. Architecture and site design are based on the base conditions and functions.

This project takes the college student fitness center as a training platform. By learning the fundamental design principles of small and medium-size span buildings, it aims to help students comprehend the logical relationship between architectural forms, space and basic structural types. Additionally, it aims to cultivate students' comprehensive ability in integrating architectural internal and external spaces, building structures and sites, as well as ensuring reasonable building lighting, ventilation, and energy conservation.

Teaching content

The current site comprises a gymnasium and a swimming pool (Lui Che-woo natatorium). The design requires that the gymnasium should be retained and the existing swimming pool should be demolished. A new fitness center for college students needs to be redesigned with a total area of approximately 4,300 m² (10% allowance). The height of the building should not exceed 24 m while considering the east-west height difference of the site. Excavation under the site should be limited to one level with a maximum depth of 4.5 m.

本科三年级

建筑设计（六）：社区文化艺术中心设计

张雷　王铠　尹航

课程类型：必修

学时学分：64 学时 / 4 学分

Undergraduate Program 3rd Year

ARCHITECTURAL DESIGN 6: DESIGN OF COMMUNITY CULTURE AND ART CENTER · ZHANG Lei, WANG Kai, YIN Hang

Type: Required Course

Study Period and Credits: 64 hours / 4 credits

课程内容

本项目拟在百子亭风貌区基地处新建社区文化中心，总建筑面积约为 8 000 m²，项目不仅为周边居民提供文化基础设施，同时也期望成为复兴老城街区活力的文化地标。根据基地条件、功能使用进行建筑和场地设计。

设计内容

1. 演艺中心：包含 400 座的小剧场，乙级。台口尺寸为 12 m×7 m。根据设计的等级确定前厅、休息厅、观众厅、舞台等面积。观众厅主要为小型话剧及戏剧表演而设置。按 60—80 人化妆布置化妆室及服装道具室，并设 2—4 间小化妆室。要求有合理的舞台及后台布置，应设有排练厅、休息室、候场区以及道具存放间等设施，其余根据需要自定。

2. 文化中心：定位成区级综合性文化站，包括公共图书阅览室、电子阅览室、多功能厅、排练厅以及辅导培训、书画创作等功能室（不少于 8 个且每个功能室面积应不低于 30 m²）。

3. 配套商业：包含社区商业以及小型文创主题商业单元。其中社区商业为不小于 200 m² 超市一处，文创主题商业单元面积为 60—200 m²。

4. 其他：变电间、配电间、空调机房、售票、办公、厕所等服务设施根据相关设计规范确定，各个功能区可单独设置，也可统一考虑。地上不考虑机动车停车配建，街区地下统一解决，但需要根据建筑功能面积计算数量。

Course content

The project plans to build a new community culture and art center at the base of Baiziting historic area, with a total construction area of about 8,000 m². The project not only serves the cultural infrastructure of surrounding residents, but also hopes to become a cultural landmark to revive the vitality of the old city. Design the building and site according to the base conditions and functional use.

Design Content

1. Performing arts center: It contains 400-seats small theatre of Class B. The size of the proscenium is 12 m × 7 m. The area of front hall, lounge, auditorium and stage shall be determined according to the design level.

2. Cultural center: It is positioned as a district-level comprehensive cultural station, including a public reading room, an electronic reading room, a multi-functional hall, a rehearsal hall, counseling and training classroom, calligraphy and painting space and other functional rooms (no less than 8 and the area of each functional room shall not be less than 30 m²).

3. Supporting business: It includes community business and small cultural and creative theme business units. Among them, the community business is a supermarket with an area of no less than 200 m², and the area of cultural and creative theme business unit is 60–200 m².

4. Others: Service facilities such as substation rooms, power distribution rooms, air conditioning rooms, ticket offices, offices and toilets are determined according to relevant design specifications.

本科四年级

建筑设计（七）：高层办公楼设计

王铠　尹航

课程类型：必修

学时学分：64 学时 / 4 学分

Undergraduate Program 4th Year

ARCHITECTURAL DESIGN 7: DESIGN OF HIGH-RISE OFFICE BUILDINGS

· WANG Kai, YIN Hang

Type: Required Course

Study Period and Credits: 64 hours / 4 credits

教学目标

本次课程设计首先希望学生了解当代建筑设计行业中高层建筑的基本特点，研究当代高层建筑的设计策略，了解高层建筑涉及的相关规范与知识，提高综合分析及解决问题的能力。其次希望学生能主动将建筑与生态、景观等环境要素有机结合，研究环境，在建筑方案中预设某种设计策略，创造出从概念、策略到解决方案一体的高层建筑设计。

教学内容

本次课程的场地位于南京市高新区南京软件园的中部核心区位置。地块面积约 1.7 hm²，要求设计一处集生产办公、人才公寓、商业配套、设备用房等于一体的复合功能建筑。

学生们须调研相关案例，并自主策划本地块的具体功能。具体面积比例在合理的基础上灵活自定。地块需考虑周边地块与交通等情况，合理组织车行、人行流线，按不同功能设置出入口集散空间；合理组织场地交通与周边道路关系，合理设置停车场。停车数量（机动车与非机动车）应尽量满足地方法规要求，地下停车场出入口大于 2 个。

Teaching objectives

The design of this course firstly hopes that students will understand the basic characteristics of high-rise buildings in the contemporary architectural design industry, study the design strategies of contemporary high-rise buildings, understand the relevant specifications and knowledge involved in high-rise buildings, and improve their ability to comprehensively analyze and solve problems. Secondly, we also hope that students can take the initiative to organically combine architecture with environmental elements such as ecology and landscape, study the environment, and preset certain design strategies in the architectural plan to create a high-rise building design that integrates concepts, strategies, and solutions.

Teaching content

The site for this course is located in the central core area of Nanjing Software Park in the High-tech Zone of Nanjing City. The plot area is about 1.7 hm², and it is required to design a composite functional building integrating production offices, talent apartments, commercial facilities, equipment rooms and so on.

Students need to research relevant cases and independently plan the specific functions of the local area. The specific area ratio can be flexibly determined on a reasonable basis. The plot needs to consider surrounding plots and traffic conditions, rationally organize vehicle and pedestrian circulation lines, set up entrance and exit distribution spaces according to different functions; rationally organize the relationship between site traffic and surrounding roads, and rationally set up parking lots. The number of parking lots (motor vehicles and non-mobile vehicles) should try to meet the requirements of local regulations, and there should be more than 2 entrances and exits to the underground parking lot.

本科四年级
建筑设计（八）：城市设计
童滋雨　唐莲
课程类型：必修
学时学分：64学时 / 4学分

Undergraduate Program 4th Year
ARCHITECTURAL DESIGN 8: URBAN DESIGN · TONG Ziyu, TANG Lian
Type: Required Course
Study Period and Credits: 64 hours / 4 credits

课程内容
计算化城市设计

教学目标
中国的城市发展已经逐渐从增量扩张转向存量更新。通过对城市建成环境的更新改造而提升环境性能和质量，将成为城市建设的新热点和新常态。与此同时，5G、物联网、无人驾驶等技术的发展又给城市环境的使用方式带来了新的变化。如何在城市更新设计中拓展建筑设计的边界也就成为新的挑战。

城市更新不但需要对建成环境本身有更充分的认知，也要对其中的人流、车流乃至水流、气流等等各种动态的活动有正确的认知。从设计上来说，这也大大提高了设计者所面临的问题的复杂性，仅靠个人的直观感受和形式操作难以保证设计的合理性。而借助空间分析、数据统计、算法设计等数字技术，我们可以更好地认知城市形态的特征，理解城市运行的规则，并预测城市未来的发展。通过规则和算法来计算生成城市也是对城市设计思维范式的重要突破。

因此，本次设计将针对这些发展趋势，以城市街巷空间为研究对象，通过思考和推演探索其更新改造的可能性。通过本次设计，学生们可以理解城市设计的相关理论和方法，掌握分析城市形态和创造更好城市环境质量的方法。

Course content
Computational Urban Design

Teaching objectives
China's urban development has gradually shifted from incremental expansion to stock renewal. Improving environmental performance and quality through the renewal and transformation of urban built environment will become a new hot-spot and new normal of urban construction. At the same time, the development of 5G, Internet of Things, unmanned driving and other technologies has brought new changes to the use of urban environment. How to expand the boundary of architectural design in urban renewal design has become a new challenge.

Urban renewal needs not only a better understanding of the built environment itself, but also a correct understanding of pedestrian and vehicle flows, water and air currents and other dynamic activities. In terms of design, it also greatly improves the complexity of the problems faced by designers. It is difficult to ensure the rationality of design only by personal intuitive feeling and formal operations. With the help of various digital technologies such as spatial analysis, data statistics and algorithm design, we can better understand the characteristics of the urban form, understand the rules of the urban operation, and predict the future development of the city. Calculating and generating cities through rules and algorithms is also an important breakthrough in the thinking paradigm of urban design.

Therefore, this design will aim at these development trends, take the urban street space as the research object, and explore the possibility of its renewal and transformation through thinking and deduction. Through this design, students can understand the relevant theories and methods of urban design, and master the methods of analyzing the urban form and creating better urban environmental quality.

本科四年级
本科毕业设计
华晓宁　赵潇欣　梁宇舒　施珊珊　史文娟
童滋雨　王洁琼　王铠　吴蔚　尹航
课程类型：必修
学时学分：1学期 / 10学分

Undergraduate Program 4th Year
GRUDUATION PROJECT
· HUA Xiaoning, ZHAO Xiaoxin, LIANG Yushu, SHI Shanshan, SHI Wenjuan, TONG Ziyu, WANG Jieqiong, WANG Kai, WU Wei, YIN Hang
Type: Required Course
Study Period and Credits: 1 term /10 credits

设计选题
赵潇欣，澳洲现代主义建筑师作品与思想研究——罗宾·博伊德和哈里·塞德勒
梁宇舒、缪晓东，游牧木构——基于轻型可变木结构的可移动帐幕类景观装置设计与建造
施珊珊，气候变化对我国城市住宅PM2.5渗透的影响
史文娟，网师园中的小型庭园景观研究
童滋雨，基于规则和算法的设计和搭建——拓扑互锁结构
王洁琼，"老旧"与"落脚"——未来新市民"一间屋"设计
王铠，新城生活服务基础设施设计研究
吴蔚，风雨归巢——南京晨光1865旧厂房绿色医养中心改造设计
尹航，基于复杂地形的参数化景观建筑设计

Design selection
ZHAO Xiaoxin, Research on the works and ideas of Australian modernist architects—Robin Boyd & Harry Seidler
LIANG Yushu, MIAO Xiaodong, Nomadic wood structure—Design and construction of movable tent landscape installations based on lightweight variable timber structures
SHI Shanshan, Effects of climate change on PM2.5 infiltration of urban residence in China
SHI Wenjuan, Study on the small-sized garden of the Master-of-Nets Garden
TONG Ziyu, Design and construction based on rules and algorithms—Topological interlocking structure
WANG Jieqiong, Old and settled—One house design for future citizens
WANG Kai, Research on the design of new town life service infrastructure
WU Wei, Returning home from wind and rain, happy to raise at home—Nanjing Chenguang 1865 renovation design of green medical care center in old factory buildings
YIN Hang, Parametric landscape architectural design based on complex terrain

研究生一年级

建筑设计研究（一）：基本设计

傅筱

课程类型：必修

学时学分：40学时 / 2学分

Postgraduate Program 1st Year

ARCHITECTURAL DESIGN RESEARCH 1: BASIC DESIGN · FU Xiao

Type: Required Course

Study Period and Credits: 40 hours / 2 credits

教学目标

课程从"场地、功能、行为、结构、经济性"等建筑的基本问题出发，通过 宅基地住宅设计训练学生对建筑逻辑性的认知，并让学生理解有品质的建筑是以基本问题为基础的整合设计。

设计要求

1）基地位于南京郊区某村。使用者设定为一对夫妇，他们有一个4岁儿子。住宅兼做丈夫的个人独立工作室，妻子为全职太太。工作室性质由设计者自定。

2）建筑用地边界不得超出原有宅基地范围（出入口踏步除外），边界位置也不得随便移动。一层悬挑的雨篷、二层悬挑的阳台（凸窗）可适当超出用地边界。

3）建筑层数2层，室内外高差0.45 m。

4）室外停小汽车2—3辆（不考虑室内停车）。

5）须进行场地布置设计

Teaching objectives

The course starts from the basic issues of architecture such as "site, function, behavior, structure, and economy". Through homestead residential design, it trains students to understand the logic of architecture and allows students to understand that quality architecture is based on basic issues.

Design requirements

1) The base is located in a village in the suburbs of Nanjing. The users are configured as a couple with a 4-year-old son. The house also serves as the husband's personal independent studio, and the wife is a full-time housewife. The nature of the studio is determined by the designer.

2) The boundary of the building land shall not exceed the scope of the original homestead (except for the entrance and exit steps), and the boundary position shall not be moved at will. The cantilevered awning on the first floor and the cantilevered balcony (bay window) on the second floor can appropriately exceed the land boundary.

3) The building has 2 floors and the indoor-outdoor height difference is 0.45 m.

4) Park 2-3 cars outdoors (indoor parking is not considered).

5) Site layout design is required.

研究生一年级

建筑设计研究（一）：基本设计

冷天

课程类型：必修

学时学分：40学时 / 2学分

Postgraduate Program 1st Year

ARCHITECTURAL DESIGN RESEARCH 1: BASIC DESIGN · LENG Tian

Type: Required Course

Study Period and Credits: 40 hours / 2 credits

教学目标

当下中国的城市建设发展，业已从"增量开发"走向"存量更新"。其中，文物建筑、历史建筑、文化遗产等大量历史性建筑，面临着如何在保存其特有历史文化价值的前提下，充分活化利用其原有空间（内部、外部）的难题。本课程希望引导学生通过一个真实的案例，直面上述的双重矛盾，理解设计的逻辑性与综合性，在历史和现实之间取得平衡，并通过一个复有创造性的设计来激发空间的活力。

设计内容

1）首都水厂清凉山水库旧址

问题与设计任务：地下蓄水池主体承重结构经常年水体腐蚀受到了不同程度的损伤与破坏，须在适宜修缮手段的基础上，对水库蓄水池闲置的空间进行活化利用，挖掘并显现其独特的历史价值和地位。

2）东南楼

问题与设计任务：东南楼将称为南京大学医学院新的公共教学空间，设计须在不破坏其历史建筑外貌的条件下提升空间品质，充分利用闲置的屋架层空间，彰显其历史文化价值。

Teaching objectives

At present, urban construction and development in China have shifted from "incremental development" to "stock renewal". Among them, a large number of historical buildings such as cultural relic buildings, historical buildings, cultural heritage, etc., are faced with the problem of how to fully activate and utilize their original spaces (internal and external) while preserving their unique historical and cultural value. This course aims to guide students through a real case to confront the above-mentioned dual contradictions, to understand the logic and comprehensiveness of design, to strike a balance between history and reality, and to stimulate the vitality of space through a creative design.

Design content

1) The former site of Qingliangshan Reservoir of Capital Water Plant

Problems and design tasks: The main load-bearing structure of the underground reservoir has suffered varying degrees of damage and destruction caused by long-term water body corrosion. It is necessary to activate and utilize the idle space of the reservoir on the basis of suitable repair methods, excavate and reveal its unique historical value and status.

2) Southeast Building

Problems and design tasks: Southeast Building will become a new public teaching space for the Medical School of Nanjing University, and the design is necessary to improve the quality of the space without destroying the appearance of its historical building, make full use of the idle attic space and highlight his historical and cultural value.

研究生一年级
建筑设计研究（一）：基本设计
潘幼建　万军杰
课程类型：必修
学时学分：40 学时 / 2 学分

Postgraduate Program 1st Year
ARCHITECTURAL DESIGN RESEARCH 1: BASIC DESIGN · PAN Youjian, WAN Junjie
Type: Required Course
Study Period and Credits: 40 hours / 2 credits

设计任务
设计课程从教师对案例的分享开始，以建立学生对概念与结构的基础意识。学生观察和调研自己所在的南京大学浦口校区，从几个备选场地中挑选一个，发现问题，依据功能主题拟定任务书，提出空间概念和结构策略，回应场地，应对问题。

结构形式来源于学生对于案例的分析（可与教师分享案例有重叠），通过分析理解结构原理，及结构与概念和空间的整体关系，以案例为基础做概念和结构的抽象和延伸，将其运用到自己的设计当中。

规模：不对建筑规模做具体的要求。但要注重建筑与周边环境的关系。

功能：依据功能主题，基于场地和空间规划自主确定项目功能构成及各部分面积，功能宜复合且具有一定的差异性。

Design tasks
The design course starts with teachers sharing cases to build students' basic awareness of concepts and structures. Students observed and investigated the Pukou Campus of Nanjing University where they were located, selected one of several alternative sites, discovered problems, formulated a task statement based on functional themes, proposed spatial concepts and structural strategies, responded to the site, and responded to the problems.

The structural form comes from students' analysis of case studies (which may overlap with the cases shared by the instructor). Through analysis, they understand the structural principles and the overall relationship between structure, concept and space, abstract and extend concepts and structures based on case studies and apply them to their own designs.

Scale: There are no specific requirements for building scale. But we should pay attention to the relationship between the building and the surrounding environment.

Function: According to the functional theme, the functional composition of the project and the area of each part are determined independently based on the site and space planning. The functions should be composite and have certain differences.

研究生一年级
建筑设计研究（一）：概念设计
鲁安东
课程类型：必修
学时学分：40 学时 / 2 学分

Postgraduate Program 1st Year
ARCHITECTURAL DESIGN RESEARCH 1: CONCEPTUAL DESIGN · LU Andong
Type: Required Course
Study Period and Credits: 40 hours / 2 credits

课程内容
本学期概念设计课程以 2022 深圳双年展为背景，聚焦"生息"这一概念，以新的认知起点重新审视建筑学科的相关性，并对急迫的城市问题做出正面回应。生息是共生的价值观，也是绵延的规律性。2022 深圳双年展以"城市生息"为主题："城市生息"秉持的是多样的自然要素的和谐共生、生生不息；"城市生息"追求的是通过对生活方式的塑造，实现人类繁衍生息；"城市生息"探索的是循环与平衡的发展之道，通过城市创新与人的行动，共同走向未来。当前城市发展面临着大转型，无论是从低碳可持续角度，还是从存量更新或者收缩城市的角度，都要求我们转向以生命为中心，更加精明、适变和包容的设计。学生将二人一组，首先构建一种"生息"类型，为其建立概念图谱，并且找到其与当代人文、技术的关联，进而加以应用，并设计一个原型性的实验方案。

Course content
This semester's conceptual design course focuses on the concept of "habitat" against the backdrop of 2022 Shenzhen Biennale, re-examining the relevance of the architectural discipline from a new cognitive starting point and responding positively to pressing urban issues. Habitat is the value of symbiosis and the regularity of continuity. 2022 Shenzhen Biennale takes "urban habitat" as its theme: "urban habitat" upholds the harmonious coexistence of diverse natural elements and endless growth; "urban habitat" pursues the realisation of the human condition through the shaping of lifestyles; "urban habitat" explores the way of cyclic and balanced development, and moves towards the future together through urban innovation and human action. Currently, urban development is facing a major transformation, whether from the perspective of low-carbon sustainability, stock renewal or shrinking cities, we are required to shift to life-centred, more astute, adaptable and inclusive design. Students will work in pairs to first construct a typology of "habitat", establish a conceptual map for it, find its connections with contemporary humanities and technology, and then apply it and design a prototype experimental scheme.

研究生一年级

建筑设计研究（一）：概念设计

窦平平

课程类型：必修

学时学分：40 学时 / 2 学分

Postgraduate Program 1st Year

ARCHITECTURAL DESIGN RESEARCH 1: CONCEPTUAL DESIGN · DOU Pingping

Type: Required Course

Study Period and Credits: 40 hours / 2 credits

题目：可感知的共生

课程内容

本课题探索如何在高度城市化、人工化的环境中与自然要素互惠共生，如何通过营造建筑环境在人类特质与自然要素之间建立隐秘又愉悦的联系。

人类天生渴望某种与自然世界有关的信息，这是植根于人类基因之中的对自然的依赖性情感反应。与自然世界的联系与大脑的快乐和痛苦中心高度相关。

共生理论营造了一种动态情境，使得两种或多种异质元素可以在保持张力的状态下共同依存，彼此支撑。

互利是共生设计的核心目标。

共生建筑是复合的、生发的，在异质中共存。

共生建筑是延续的、动态的，在变化中获得稳定。

设计要求

课程要求提出一种共生建筑的概念，及其相应的空间模式、交互机制、感知方式、关键细节。寻找和呈现环境中隐匿的能量流动与制约关系。探索不以人类为中心的且有益于人类的设计。

Title: Sensible Symbiosis

Course content

This project explores how to live in symbiosis with natural elements in a highly urbanised and artificial environment, and how to create a hidden yet pleasurable connection between human traits and natural elements through the creation of a built environment.

Human beings have an innate desire for some kind of information about the natural world, and this is a dependent emotional response to nature that is rooted in our genes. The connection with the natural world is highly correlated with the brain's centers of pleasure and pain.

Symbiosis theory creates a dynamic situation in which two or more heterogeneous elements can co-depend and support each other while maintaining tension.

Mutual benefit is the central goal of symbiotic design.

Symbiotic architecture is composite and generative, and coexists with heterogeneity.

Symbiotic architecture is continuous and dynamic, and achieves stability through change.

Course requirements

The course requires the presentation of a concept of symbiotic architecture and its corresponding spatial patterns, mechanisms of interaction, modes of perception, and key details. Seek and present the hidden energy flows and constraints in the environment. Explore designs that are not anthropocentric but beneficial to humans.

研究生一年级

建筑设计研究（二）：综合设计

程向阳

课程类型：必修

学时学分：40 学时 / 2 学分

Postgraduate Program 1st Year

ARCHITECTURAL DESIGN RESEARCH 2: COMPREHENSIVE DESIGN · CHENG Xiangyang

Type: Required Course

Study Period and Credits: 40 hours / 2 credits

场地概况

中国科学院南京土壤研究所位于南京市玄武区北京东路71 号，成立于 1953 年，其前身为 1930 年创立的中央地质调查所土壤研究室。中国科学院南京地理与湖泊研究所位于南京市玄武区北京东路 73 号，其前身是 1940 年 8 月成立的中国地理研究所，1988 年 1 月改名为中国科学院南京地理与湖泊研究所并沿用至今，是中国唯一以湖泊—流域系统为主要研究对象的地理研究所。两个研究所共用一个大院空间，没有明确的物理边界分割，总占地面积为 34 760 m²。

设计任务

1. 基于城市尺度分析，研究如何对大院搬迁后的空间进行再定义，让后续空间的策划与运营适应新的社会环境和社会需求，并建立可持续的发展模式。

2. 以既存空间作为研究主线，分析空间在都市中潜在的特质，进行系列的空间实验。

3. 以空间作为设计实践的对象，架构适配新的内容，激发空间的创造性和价值。

Site overview

Nanjing Institute of Soil Science, Chinese Academy of Sciences, is located at No. 71 Beijing East Road, Xuanwu District, Nanjing City. It was founded in 1953 and originated from the Soil Research Office of the Central Geological Survey Institute, which was founded in 1930. Nanjing Institute of Geography and Limnology, Chinese Academy of Sciences, is located at No. 73 Beijing East Road, Xuanwu District, Nanjing City. Its predecessor was the Institute of Geography of China, established in August 1940, and it was renamed Nanjing Institute of Geography and Limnology, Chinese Academy of Sciences in January 1988, which has been in use ever since. It is the only geographical research institute in China with the lake-basin system as the main object of research. The above two institutes share the same compound space without a clear physical boundary division, covering a total area of 34,760 m².

Design tasks

1. Based on urban scale analysis, research on how to redefine the space after the relocation of the compound, let the planning and operation of the subsequent space adapt to the new social environment and social demand, and establish a sustainable development mode.

2. Taking the existing space as the main line of research, analyse the potential qualities of space in the city, and conduct a series of spatial experiments.

3. Taking space as the object of design practice, structure and accommodate new content, stimulate the creativity and value of the space.

研究生一年级

建筑设计研究（二）：综合设计
周凌
课程类型：必修
学时学分：40学时 / 2学分

Postgraduate Program 1st Year

ARCHITECTURAL DESIGN RESEARCH 2: COMPREHENSIVE DESIGN · ZHOU Ling
Type: Required Course
Study Period and Credits: 40 hours / 2 credits

题目：都市再生

课程内容

基地一：以南京河西奥体地块为例，针对前期城市化过程中遗留下来的不足与部分功能性过时等问题，通过对经济、社会、文化、环境、交通、治理等的分析、研究，提出具有创造性的解决方案，达到地区都市再生与再城市化的目的。

基地二：以南京大学鼓楼校区周边地块与街道为例，针对人群、商业、场所要素，进行重构再生，以适应新的城市发展需求。

Title: Urban Regeneration

Research content

Base 1: Taking Nanjing Hexi Olympic Sports Center plot as an example, targeting the problems of deficiencie and partial functional obsolescence left over from the pre urbanisation process, the study proposes creative solution through analyses and researches on economic, socia cultural, environmental, traffic and governance etc., so a to achieve the purpose of regional urban regeneration an re-urbanization.

Base 2: Taking the plots and streets surrounding th Gulou Campus of Nanjing University as an example, th study focuses on the reconstruction and regeneration population, commerce, and place elements to adapt to th new needs of urban development.

研究生一年级

建筑设计研究（二）：综合设计
金鑫
课程类型：必修
学时学分：40学时 / 2学分

Postgraduate Program 1st Year

ARCHITECTURAL DESIGN RESEARCH 2: COMPREHENSIVE DESIGN · JIN Xin
Type: Required Course
Study Period and Credits: 40 hours / 2 credits

研究内容

1. 尺度转换

工业建筑、构筑物和场地等通常具有远超人体尺度的巨大体量，并容纳大量复杂的机器和设备的运作。这样的物质环境，具有机器化、非人性的尺度和空间，难以与常人的生活、工作等活动相关联。以空间的再生作为设计研究的核心，意味着将工业建筑的巨大空间转向民用、公共的空间。

而工业建筑巨大的空间尺度和坚固的结构体系，提供了重新组织交通流线的可能。这同样需要做相应的空间尺度转换研究，以汽车和车行的尺度作为基本单元来研究工业建筑的空间适应性。

2. 程序重置

在原为满足生产工艺流程要求而设置的工业建筑空间及其组织关系中，重新置入符合城市生活需求的新的程序与功能，合理安排新的活动内容。

3. 结构重组

在工业建筑的改造中，为了满足空间再利用的需求，可对工业建筑既有结构体系进行改变或重组。新置入的结构体与既有工业建筑结构体系可能形成多种空间位置和受力关系。

Research content

1. Scale conversion

Industrial buildings, structures, and sites usually hav huge volumes that far exceed the scale of the huma body, and accommodate the operation of a large numb of complex machines and equipment. Such a materi environment has a mechanized and inhuman scale an space, and is difficult to relate to ordinary people's lif work and other activities. Taking the regeneration of spac as the core of design research means turning the hug space of industrial buildings into civil and public spaces.

The huge spatial scale and solid structural system industrial buildings provide the possibility to reorgani traffic flow. This also requires corresponding research spatial scale conversion, using the scale of automobil and vehicular traffic as the basic unit to study the spat adaptability of industrial buildings.

2. Program reset

In the industrial building space and its organizatio relationship that are originally set up to meet th requirements of the production process, new procedur and functions that meet the needs of urban life are r installed, and new activities are reasonably arranged.

3. Restructuring

In the transformation of industrial buildings, in ord to meet the needs of space reuse, the existing structu system of industrial buildings can be changed or reorganize The newly placed structure may form a variety of spat positions and mechanical relationships with the existi industrial building structure system.

研究生一年级

建筑设计研究（二）：城市设计

华晓宁

课程类型：必修

学时学分：40学时 / 2学分

Postgraduate Program 1st Year
ARCHITECTURAL DESIGN RESEARCH 2: URBAN DESIGN · HUA Xiaoning
Type: Required Course
Study Period and Credits: 40 hours / 2 credits

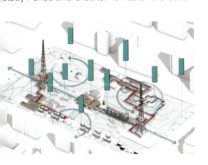

研究课题

信息·机器·辖域——信息基础设施的空间政治学（"国民政府中央广播电台"旧址再生）

课程议题

基础设施日益成为当代城市研究与实践的重要主题。以往基础设施仅仅被视作市政工程的专业领域，遵循工具理性，被传统建筑学忽视多年。然而基础设施内在具有的独特空间潜力，能够成为重塑空间状态和空间关系的触媒。

作为传送、分发信息的机器，信息基础设施的"源""流""域"都具有天然的空间性，赋予空间不同的属性特征。本课题关注城市中的信息基础设施，解读和反思其引发的空间分异，并在当代语境中对其重新定义，创造新的空间关联，从而推动对城市空间的重构。

设计要求

对"国民政府中央广播电台"旧址及其周边地段进行深入调研，探寻该基础设施在城市发展中的潜力，构想该场址的未来愿景，提出场址的再生策略，并选择重点建筑进行改造设计。

Research topic

Information · Machine · Territory—The Spatial Politics of Information Infrastructure (Regeneration of the Former Site of the "Nationalist Government's Central Broadcasting Station")

Course topic

Infrastructure has increasingly become an important topic in contemporary urban research and practice. In the past, infrastructure was merely seen as a professional field of municipal engineering, following instrumental rationality, and neglected for many years by traditional architecture. However, infrastructure inherently possesses unique spatial potential, which can serve as a catalyst for reshaping spatial conditions and spatial relationships.

As a machine for transmitting and distributing information, the "source," "flow," and "domain" of information infrastructure all have inherent spatiality, endowing space with different attribute characteristics. This study focuses on information infrastructure in cities, interpreting and reflecting on the spatial differentiation it triggers, and redefining it in the contemporary context, creating new spatial relationships to drive the reconstruction of urban space.

Design requirements

Conduct in-depth research on the former site of the "Nationalist Government's Central Broadcasting Station" and its surrounding areas, explore the potential of this infrastructure in urban development, envision the future vision of the site, propose strategies for its revitalization, and select key buildings for renovation and design.

研究生一年级

建筑设计研究（二）：城市设计

胡友培

课程类型：必修

学时学分：40学时 / 2学分

Postgraduate Program 1st Year
ARCHITECTURAL DESIGN RESEARCH 2: URBAN DESIGN · HU Youpei
Type: Required Course
Study Period and Credits: 40 hours / 2 credits

研究课题

都市区重构——一个关于开放城市结构的设计试验

背景与议题

当下中国城市的发展进入都市区阶段，在稠密的传统中心城区之外是尺度巨大的都市区域。在城市存量发展与更新背景下，这很可能是未来城市化深耕的主战场。

相比于成熟的传统城区，都市区域的特征之一是快速而野蛮的生长导致发展形态碎片化，各种异质的要素毫无征兆地碰撞在一起，产生出混乱但生机勃勃的都市景观，即所谓的城乡接合部。针对该特征，应考虑如何在碎片化的现状中重构出某种都市空间秩序或形态结构，为松散粗放的都市环境注入某种都市性。

都市区域的特征之二是变动不定以及不确定的未来。这导致无法运用蓝图式规划思维加以应对，任何构图式的总体规划都注定失效。针对该特征，应考虑是否存在一种动态的、开放的城市结构与形态，以容纳都市区不确定的未来与动态。

都市区域的特征之三是粗放圈地开发已经基本成型，全盘推翻毫无可能，只能以城市更新的方式修修补补，这无疑为秩序的重构带来极大的难度。针对该特征，应考虑如何以最低的成本，在保持全局视野的基础上，选取战略性点位进行更新与重构。

Research topic

Metropolitan Area Restructuring—A Design Experiment on Open Urban Structure

Background and topic

The current development of Chinese cities has entered the metropolitan area stage. Beyond the dense traditional central city areas, there are vast metropolitan areas. In the context of urban stock development and renewal, this is likely to be the main battlefield for future urbanization.

Compared to mature traditional urban areas, the first characteristic of metropolitan areas is rapid and brutal growth, leading to fragmented development forms, various heterogeneous elements, and sudden collisions that create chaotic but vibrant urban landscapes, known as urban periphery. Aiming at this characteristic, consideration should be taken to how to reconstruct some form of urban spatial order or morphological structure in the fragmented current situation, injecting a sense of urbanity into the loose and extensive urban environment.

The second characteristic of metropolitan areas is their variability and uncertain future. This makes it impossible to respond with blueprint-style planning thinking, as any master planning approach is destined to fail. Aiming at this characteristic, it should be considered whether there exists a dynamic and open urban structure and morphology that can accommodate the uncertain future and dynamics of metropolitan areas.

The third characteristic of metropolitan areas is that extensive land development has already taken shape and cannot be completely overturned. It can only be repaired and patched up through urban renewal, which undoubtedly brings great difficulty to the reconstruction of order. Aiming at this characteristic, consideration should be taken to how to select strategic points for renewal and reconstruction with the lowest cost while maintaining a comprehensive vision.

研究生国际教学工作坊
回收边缘：作为城市设备的校园
保拉·佩莱格里尼
课程类型：选修
学时学分：18学时 / 1学分

Postgraduate International Design Studio
RECYCLING THE EDGES: THE CAMPUS AS AN URBAN DEVICE ·
Paola PELLEGRINI
Type: Elective Course
Study Period and Credits: 18 hours / 1 credit

背景介绍
可持续再生——尽可能减少对环境的负面影响——是一个挑战，它需要仔细观察和创造力地寻找新的机会和设计方式，以适应现有城市。工作坊的练习以大学校园及其周边的城市空间为目标。学生必须观察校园边缘的空间——校园内外、现有建筑和开放空间。学生必须设想如何改变边缘位置，增加新的用途和建筑结构，使空间更加高效。挑战在于理解校园如何成为城市设备，即如何提供空间和结构，以供校园学生和职员以及市民使用，特别是周边区域的居民。

课程限定与要求
学生必须确定一个小空间，了解为什么它未被使用，定义新用途和用户的项目（将用于什么？谁将使用这个空间？），起草设计以提高其性能。城市设计尤其注重开放的公共空间，因此设计中也必须包含一个开放区域。

Background description
Sustainable regeneration—one that reduces as much as possible the negative impacts on the environment—is a challenge that requires careful observation and creativity for finding new opportunities and design approaches in the existing city. The exercise of the workshop targets the university campus and its edges with the surrounding urban spaces. The students must observe the spaces along the edges of the campus—both inside and outside the campus, both the existing buildings and open spaces. Students must imagine where and how the edges can be transformed and new uses and building structures can be added to make the space more efficient. The challenge is to understand how the campus can become an urban device, meaning how it can offer spaces and structures for uses and activities that can benefit both the campus students and staff and the citizens, in particular the residents of the surrounding areas.

Course restrictions and requirements
Students must identify a small space, understand why it is under-used, define a program of new uses and users (what is it going to be used for? Who will use the space?) draft a design to increase its performance. Urban design pays particular attention to open public space so there must be also an open area in the design.

研究生国际教学工作坊
可持续建筑设计
何塞·阿莫多瓦尔 唐莲
课程类型：选修
学时学分：18学时 / 1学分

Postgraduate International Design Studio
SUSTAINABLE ARCHITECTURAL DESIGN · Jose ALMODOVAR, TANG Lian
Type: Elective Course
Study Period and Credits: 18 hours / 1 credit

教学目标
本课程介绍学生如何设计可持续发展建筑和使用不同的环境设计工具。通过综合分析和计算机模拟物理环境、建筑类型、取向、内部配置、围护结构的影响，为学习和探索新的设计概念和策略提供了跨学科和跨文化的环境。环境设计的概念和方法将指导整个设计过程。学生将考虑当地和全球的文化价值观，提出高能源效率和环境质量的创新建议。该课程包括两个部分：系列讲座和设计工作室，强调做中学。

课程要求
要求学生根据课程计划展示他们的工作材料。最终演示应包含A0面板，展示不同设计方案的分析方案、计算机模拟图、比较表、计划和部分、最终方案的透视 / 轴测法、关于设计过程和概念讨论的团队声明、项目能源和环境绩效的数据摘要。

Teaching objectives
This course introduces students to how to design sustainable buildings and use different environmental design tools. It provides a cross-disciplinary and cross-cultural environment for learning and exploring new design concepts and strategies through integrated analysis and computer simulations of effects of physical environment building typology, orientation, internal configuration thermal envelope. The concepts and methods of environmental design will guide the whole design process. Students will take into account local and global cultural values to make innovative proposals of high energy efficiency and environmental quality. The course includes two parts: lecture series and design studio that emphasis on learning by doing.

Course requirements
Students are required to present their working material according to the course program. The final presentation should contain A0 panels showing analytical schemes of different design options, computer simulation diagrams comparative tables, plans and sections, perspective axonometric of final proposal, team statement about the design process and conceptual discussion, data summary of the energy and environmental performance of the project.

建筑理论课程
ARCHITECTURAL THEORY COURSES

本科一年级
建筑通史 · 王骏阳
课程类型：平台（工科试验班）
学时 / 学分：36 学时 / 2 学分

Undergraduate Program 1st Year
**GENERAL HISTORY OF ARCHITECTURE ·
WANG Junyang**
Type: Platform (Engineering Experimental Class)
Study Period and Credits: 36 hours / 2 credits

课程简介
作为一门建筑与城规实验班本科一年级学科基础认知课程和外专业通识选修课程，建筑通史并不仅是传统意义上从古代到当代的逐个历史分期的"通"，也是打破中外建筑史和中西建筑史学科隔阂的"通"，是古今中外的"通"，是建筑史与建筑学概论的"通"，是建筑学专业认知与人类文明和文化历史的"通"。本课程的教学旨在培养学生立足本土胸怀世界的人文情怀和广博的专业知识技能。

课程内容
第一讲　走出建筑与建筑物的悖论
第二讲　居住与建筑的起源与发展（一）
第三讲　居住与建筑的起源与发展（二）
第四讲　神秘的人类早期文明
第五讲　融合与冲突中的西亚与伊斯兰建筑
第六讲　古希腊和古罗马建筑
第七讲　欧洲中世纪建筑
第八讲　意大利文艺复兴建筑及后续
第九讲　现代建筑
第十讲　密斯·凡·德·罗、勒·柯布西耶、阿道夫·路斯：现代建筑的三种道路
第十一讲　后现代建筑
第十二讲　日本建筑
第十三讲　世界建筑史中的中国建筑与西村大院
第十四讲　结构、空间、形式

Course introduction
As a first-year undergraduate suject-based course of Architecture and Urban Planning Experimental class and a general elective course for non-major students, General History of Architecture is not only a traditional "general" course from ancient times to contemporary times, but also a "general" course that breaks down the disciplinary boundaries between Chinese and foreign architectural history and Chinese and Western architectural history, a "general" course between ancient and modern times, and a "general" course between the history of architecture and introduction to architecture, and a "generalization" of the professional knowledge of architecture and the history of human civilization and culture. The aim of this course is to cultivate students humanistic sentiment of having a global vision while being rooted in the local community, as well as broad professional knowledge and skills.

Course content
Lecture 1 Out of the Paradox of Architecture and Buildings
Lecture 2 The Origin and Development of Habitat and Architecture (I)
Lecture 3 The Origin and Development of Habitat and Architecture (II)
Lecture 4 The Mysterious Early Human Civilization
Lecture 5 West Asian and Islamic Architecture in the Context of Integration and Conflict
Lecture 6 Ancient Greek and Roman Architecture
Lecture 7 European Medieval Architecture
Lecture 8 Italian Renaissance Architecture and Its Aftermath
Lecture 9 Modern Architecture
Lecture 10 Ludwig Mies van der Rohe, Le Corbusier, Adolf Loos: Three Paths of Modern Architecture
Lecture 11 Postmodern Architecture
Lecture 12 Japanese Architecture
Lecture 13 Chinese Architecture and the West Village Basis Yard in the History of World Architecture
Lecture 14 Structure, Space, Form

本科二年级 / 本科三年级
建筑设计基本原理 · 周凌
课程类型：必修
学时 / 学分：36 学时 / 2 学分

Undergraduate Program 2nd Year/ 3rd Year
**BASIC THEORY OF ARCHITECTURAL DESIGN
· ZHOU Ling**
Type: Required Course
Study Period and Credits: 36 hours / 2 credits

教学目标
本课程是建筑学专业本科生的专业基础理论课程。本课程的任务主要是介绍建筑设计中形式与类型的基本原理。形式原理包含历史上各个时期的设计原则，类型原理讨论不同类型建筑的设计原理。

课程要求
1. 讲授大纲的重点内容；
2. 通过分析实例启迪学生的思维，加深学生对有关理论及其应用、工程实例等内容的理解；
3. 通过对实例的讨论，引导学生运用所学的专业理论知识，分析、解决实际问题。

课程内容
1. 形式与类型概述
2. 古典建筑形式语言
3. 现代建筑形式语言
4. 当代建筑形式语言
5. 类型设计
6. 材料与建造
7. 技术与规范
8. 课程总结

Teaching objectives
This course is a basic theory course for the undergraduate students of architecture. The main purpose of this course is to introduce the basic principles of the form and type in architectural design. Form theory contains design principles in various periods of history; type theory discusses the design principles of different types of the building.

Course requirements
1. Teach the key elements of the outline;
2. Enlighten students' thinking and enhance students' understanding of the theories, its applications and project examples through analyzing examples;
3. Help students to use the professional knowledge to analyse and solve practical problems through the discussion of examples.

Course content
1. Overview of forms and types
2. Classical architecture form language
3. Modern architecture form language
4. Contemporary architecture form language
5. Type design
6. Materials and construction
7. Technology and specification
8. Course summary

本科三年级
居住建筑设计与居住区规划原理 · 冷天 刘铨
课程类型：必修
学时/学分：36学时/2学分

Undergraduate Program 3rd Year
THEORY OF HOUSING DESIGN AND RESIDENTIAL PLANNING • LENG Tian, LIU Quan
Type: Required Course
Study Period and Credits: 36 hours / 2 credits

课程内容
第一讲 课程概述
第二讲 居住建筑的演变
第三讲 套型空间的设计
第四讲 套型空间的组合与单体设计（一）
第五讲 套型空间的组合与单体设计（二）
第六讲 居住建筑的结构、设备与施工
第七讲 专题讲座：住宅的适应性，支撑体住宅
第八讲 城市规划理论概述
第九讲 现代居住区规划的发展历程
第十讲 居住区的空间组织
第十一讲 居住区的道路交通系统规划与设计
第十二讲 居住区的绿地景观系统规划与设计
第十三讲 居住区公共设施规划、竖向设计与管线综合
第十四讲 专题讲座：住宅产品开发
第十五讲 专题讲座：住宅产品设计实践
第十六讲 课程总结，考试答疑

Course content
Lecture 1 Course overview
Lecture 2 Evolution of residential buildings
Lecture 3 Design of dwelling space
Lecture 4 Dwelling space arrangement and monomer building design (1)
Lecture 5 Dwelling space arrangement and monomer building design (2)
Lecture 6 Structure, facility and construction of residential buildings
Lecture 7 Special lecture: Adaptability of housing, SI housing
Lecture 8 An overview of urban planning theory
Lecture 9 The development history of modern residential area planning
Lecture 10 Spatial organization of residential area
Lecture 11 Traffic system planning and design of residential area
Lecture 12 Green landscape system planning and design of residential area
Lecture 13 Public facilities planning, vertical design and pipeline integration of residential area
Lecture 14 Special lecture: Residential product development
Lecture 15 Special lecture: Residential product design practice
Lecture 16 Course summary and exam Q&A

本科四年级
建筑设计行业知识与创新实践 · 周凌 梁宇舒
课程类型：选修
学时/学分：36学时/2学分

Undergraduate Program 4th Year
KNOWLEDGE AND INNOVATIVE PRACTICE IN THE ARCHITECTURAL DESIGN INDUSTRY • ZHOU Ling, LIANG Yushu
Type: Elective Course
Study Period and Credits: 36 hours / 2 credits

课程内容
建筑设计是一项应用性以及科学、艺术、技术、人文综合性很强的学科与工作。本课程以兼顾本科毕业生及研究生的本硕贯通课，同时面向校内其他专业同学的公共选修课为定位，作为南京大学校级"创新创业"立项课程。教学注重学生学习方法、创新能力、主动性思考、沟通表达能力的培养，将建筑学背景下的行业知识与创新实践相结合，旨在帮助学生拓展行业视野，了解前沿的行业知识与多元的行业发展动态，致力于建设面向院外的高质量开放课程。课程同步面向公众进行线上推广，逐步建设南京大学建筑与城市规划学院大师课程库。

Course content
Architectural design is a highly applied discipline and work with strong integration of science, art, technology and humanities. This course is designed to accommodate both undergraduate and postgraduate students in a master's program, as well as a public elective course for other majors on campus and a school level "Innovation and Entrepreneurship" course approved by Nanjing University. It focuses on the cultivation of students' learning method, innovation ability, initiative thinking, communication and expression ability and quality, and combines industry knowledge and innovative practice in the context of architecture, intending to help students develop industry awareness for students, and understand the cutting-edge industry knowledge and a diversified vision of the development of the industry, with the aim of building a high-quality open course for the outside of the university. The course will be synchronized with online promotion for the public, and gradually build a master course library for the School of Architecture and Urban Planning of Nanjing University.

研究生一年级/博士生
建筑与城市研究基础 · 鲁安东
课程类型：必修
学时/学分：16学时/1学分

Postgraduate Program 1st Year / Ph.D. student
FOUNDATIONS OF ARCHITECTURE AND URBAN STUDIES • LU Andong
Type: Required Course
Study Period and Credits: 16 hours / 1 credit

课程内容
第一讲 绪论：学术研究的基本认识和常见方法
　　1. 设计、研究、实践
　　2. 元素、路径
第二讲 研究题目
　　1. 问题、论题、命题
　　2. 初学者研究题目的选定
　　3. 如何评价研究题目
第三讲 研究计划
　　1. 选择研究方法
　　2. 设计研究路径
　　3. 制订研究计划
第四讲 文献阅读与综述
　　1. 如何阅读文献
　　2. 如何撰写文献综述
第五讲 研究成果的表达
　　1. 利用视觉工具
　　2. 考虑听众的阅读
　　3. PPT与演讲的关系
第六讲 学位论文写作的基本规范与评价
　　1. 学位论文写作的基本规范
　　2. 学位论文的评审
第七讲 优秀论文分析（规划）
第八讲 优秀论文分析（建筑）

Course content
Lecture 1 Introduction: Basic understanding and common methods of academic research
　　1. Design, research, practice
　　2. Elements, paths
Lecture 2 Research topics
　　1. Problem, thesis, proposition
　　2. Selection of research topics for beginners
　　3. How to evaluate research topics
Lecture 3 Research program
　　1. Choosing a research method
　　2. Designing a research path
　　3. Developing a research plan
Lecture 4 Literature reading and review
　　1. How to read literature
　　2. How to write a literature review
Lecture 5 Expression of research results
　　1. Utilizing visual tools
　　2. Considering the audience's reading
　　3. The relationship between PPT and presentation
Lecture 6 Basic standards and evaluation of dissertation writing
　　1. Basic standards of dissertation writing
　　2. Review of dissertations
Lecture 7 Analysis of excellent dissertations (planning)
Lecture 8 Analysis of excellent dissertation (architecture)

研究生一年级
建筑理论研究・赵辰
课程类型：必修
学时 / 学分：18 学时 / 1 学分

Postgraduate Program 1st Year
STUDIES OF ARCHITECTURAL THEORY
• ZHAO Chen
Type: Required Course
Study Period and Credits:18 hours / 1 credit

课程介绍
　　了解中、西方学者对中国建筑文化诠释的发展过程，理解新的建筑理论体系中对中国建筑文化重新诠释的必要性，学习重新诠释中国建筑文化的建筑观念与方法。

课程内容
1. 本课的总览和基础。
2. 中国建筑：西方人的诠释与西方建筑观念的改变。
3. 中国建筑：中国人的诠释以及中国建筑学术体系的建立。
4. 木结构体系：中国建构文化的意义。
5. 住宅与园林：中国人居文化的意义。
6. 宇宙观的和谐：中国城市文化的意义。
7. 讨论。

Course description
Understand the development process of Chinese and Western scholars' interpretation of Chinese architectural culture, understand the necessity of reinterpretation of Chinese architectural culture in the new architectural theory system, and learn the architectural concepts and methods of reinterpretation of Chinese architectural culture.

Course content
1. Overview and foundation of this course.
2. Chinese architecture: Western interpretation and the change of Western architectural concept.
3. Chinese architecture: Chinese interpretation and the establishment of Chinese architecture academic system.
4. Wood structure system: The significance of Chinese construction culture.
5. Residence and garden: The significance of Chinese human settlement culture.
6. Harmony of cosmology: The significance of Chinese urban culture.
7. Discussion.

建筑技术课程
ARCHITECTURAL TECHNOLOGY COURSES

本科二年级
CAAD 理论与实践（一）· 吉国华 傅筱 万军杰
课程类型：选修
学时 / 学分：36 学时 / 2 学分

Undergraduate Program 2nd Year
THEORY AND PRACTICE OF CAAD 1
• JI Guohua, FU Xiao, WAN Junjie
Type: Elective Course
Study Period and Credits: 36 hours / 2 credits

课程介绍
在现阶段的 CAD 教学中，强调建筑设计在建筑学教学中的主导地位，将计算机技术定位于绘图工具。本课程旨在帮助学生尽快并且熟练地掌握利用计算机工具进行建筑设计的表达。课程中整合了 CAD 知识、建筑制图知识以及建筑表现知识，将传统 CAD 教学中教会学生用计算机绘图的模式向教会学生用计算机绘制有形式感的建筑图的模式转变，强调准确性和表现力作为评价 CAD 学习的两个最重要指标。
本课程的具体学习内容包括：
1. 初步掌握 AutoCAD 软件和 SketchUP 软件的使用，能够熟练完成二维制图和三维建模的操作；2. 掌握建筑制图的相关知识，包括建筑投影的基本概念，平立剖面、轴测、透视和阴影的制图方法和技巧；3. 图面效果表达的技巧，包括黑白线条图和彩色图纸的表达方法和排版方法。

Course description
The core position of architectural design is emphasized in the CAD course. The computer technology is defined as a drawing tool. The course helps students learn how to make architectural presentation using computers fast and expertly. The knowledge of CAD, architectural drawing and architectural presentation are integrated in the course. The traditional mode of teaching students to draw in CAD course will be transformed into teaching students to draw architectural drawing with sense of forms. The precision and expression will be emphasized as two most important factors to estimate the teaching effect of CAD course.
The specific learning content of the course includes:
1. Use AutoCAD and SketchUP to achieve the 2D drawing and 3D modeling expertly; 2. Learn relational knowledge of architectural drawing, including basic concepts of architectural projection, drawing methods and skills of plan, elevation, section, axonometry, perspective and shadow; 3. The techniques for expressing graphic effects, including the methods of expression and layout for both black-and-white line drawings and colored drawings.

本科二年级
建筑力学 · 董萼良
课程类型：必修
学时 / 学分：32 学时 /2 学分

Undergraduate Program 2nd Year
ARCHITECTURAL MECHANICS
• DONG Eliang
Type: Required Course
Study Period and Credits: 32 hours / 2 credits

教学目标
本课程旨在培养学生求真务实、精益求精的建筑学专业态度和职业精神。通过学习本课程，使学生了解结构受力分析的基础知识；熟练掌握静力学的基本知识、平面力系物体系统平衡问题的求解方法；掌握杆件的基本变形和截面内力（轴力、剪力和弯矩）计算；掌握强度、刚度的计算和基本概念。

课程内容
一、静力学基本知识
1. 静力学公理和物体的受力分析；
2. 平面汇交力系和平面力偶系；
3. 平面力系的合成和平衡；
4. 物体系统的平衡问题；
5. 平面桁架的内力计算。
二、材料力学基础
1. 杆件变形的基本形式；
2. 轴力的概念和计算方法；
3. 剪切的概念；
4. 扭转的概念和计算方法；
5. 梁平面弯曲的概念和内力计算；
6. 梁平面弯曲时内力图的绘制；
7. 应力、应变及强度、刚度的计算和基本概念。

Teaching objectives
This course aims to cultivate students' professional attitude and professional spirit of being pragmatic and striving for excellence in architecture. Through studying this course, students will understand the basic knowledge of structural force analysis; master the basic knowledge of statics, the solution method of the plane force system object system equilibrium problem; master the knowledge of basic deformation of rods and the calculation of cross-sectional internal forces (axial force, shear force, and bending moment); master the calculation and basic concepts of strength and stiffness.

Course content
I. Basic knowledge of statics
 1. Axioms of statics and force analysis of objects;
 2. Plane convergent force system and plane couple system;
 3. Synthesis and balance of plane force system;
 4. Balance problem of object system;
 5. Calculation of internal force of plane truss.
II. Basics of material mechanics
 1. Basic forms of rod deformation;
 2. Concept and calculation method of axial force;
 3. Concept of shear;
 4. Concept and calculation method of torsion;
 5. Concept of plane bending of beam and calculation of inner force;
 6. Drawing of internal force diagram when plane bending of beam;
 7. Calculation and basic concepts of stress, strain, strength and stiffness.

本科二年级
建筑结构 · 孟宪川
课程类型：必修
学时 / 学分：32 学时 /2 学分

Undergraduate Program 2nd Year
BUILDING STRUCTURE • MENG Xianchuan
Type: Required Course
Study Period and Credits: 32 hours / 2 credits

课程介绍
本课程将重点从建筑结构的纯理论知识，转变为以结构知识辅助建筑设计的思维与能力教学，引导学生主动关注身边的结构问题，由此延伸至影响建筑设计的结构理论与计算，最后将结构问题转化为建筑设计策略，使结构知识成为其建筑实践中重要的创新立足点。
本课程旨在使学生掌握基本的建筑结构知识，并能将其转化为建筑设计的辅助方法，由此解决以往教学内容中建筑结构理论难以服务建筑设计的难题。同时初步培养学生以结构专业能力尝试挑战与应对设计中创新的形式，以及与结构工程师沟通的能力。

课程内容
本课程具体教学内容分为三阶段：
1. 结构基础 + 现有公交亭案例的结构分析；
2. 结构力学与计算 + 优秀公交亭案例的结构分析；
3. 结构设计 + 公交亭的建筑结构设计。

Course description
This course shifts the focus from the pure theoretical knowledge of building structure to the thinking and ability teaching of structural knowledge assisted architectural design, guiding students to focus the structural problems around them actively, and then extend to the structural theory and calculation that affect architectural design, and finally transform the structural problems into architectural design strategies, so that structural knowledge becomes an important innovation foothold in students' architectural practice.
This course aims to enable students to master basic building structure knowledge and transform it into an auxiliary method for architectural design, thereby solving the problem that building structure theory is difficult to serve architectural design in previous teaching content. At the same time, students are initially trained to try to challenge and respond to innovative forms in design with structural professional ability, as well as the ability to communicate with structural engineers.

Course content
The specific teaching content of this course is divided into three stages:
1. Basic knowledge of structure and structural analysis of existing bus shelter cases;
2. Structural mechanics and calculation and structural analysis of excellent bus shelter cases;
3. Structural design and building structure design of bus shelters.

本科二年级
建筑与规划技术（二）（模块）· 徐建刚 甄峰 尹海伟 祁毅 胡宏 张姗琪
课程类型：必修（实验班技术课程）
学时 / 学分：64 学时 / 4 学分

Undergraduate Program 2nd Year
ARCHITECTURE AND PLANNING TECHNOLOGY II (MODULE) · XU Jiangang, ZHEN Feng, YIN Haiwei, QI Yi, HU Hong, ZHANG Shanqi
Type: Required Course (Technical courses for experimental classes)
Study Period and Credits: 64 hours / 4 credits

课程介绍
本课程是南京大学城乡规划学科规划技术类课程的基础课程，是数字城市规划与智慧生态城市规划系列课程的重要组成部分，为国土空间规划、城市更新和城市设计等方向的学习提供基本的数据思维范式和技术方法支撑。

本课程的任务是使学生熟悉国土空间规划、城市更新和城市设计等方向中的常用数据，熟练掌握多源数据的获取、处理、分析与应用的技术流程与基本操作，并注重培养学生的综合分析能力与批判性思维素养，使学生具备正确选择和综合运用各种软件、模型与技术定量解决各类规划问题的能力。

Course description
The course is a foundational course in the planning technology category of the urban and rural planning discipline at Nanjing University. It is an important component of the series of courses on digital urban planning and smart ecological urban planning, providing basic data thinking paradigms and technical methods to support the study of national spatial planning, urban renewal, and urban design. The tasks of this course are to familiarize students with the commonly used data in the directions of national spatial planning, urban renewal, and urban design, proficiently master the technical processes and basic operations for obtaining, processing, analyzing, and applying multi-source data, and focus on cultivating students' comprehensive analytical abilities and critical thinking skills, enabling them to correctly select and comprehensively apply various software, models, and technologies to quantitatively solve various planning problems.

本科二年级
建筑与规划技术（一）（模块）· 傅筱 孟宪川 吉国华 万军杰
课程类型：必修（实验班技术课程）
学时 / 学分：64 学时 / 4 学分

Undergraduate Program 2nd Year
ARCHITECTURE AND PLANNING TECHNOLOGY I (MODULE) · FU Xiao, MENG Xianchuan, JI Guohua, WAN Junjie
Type: Required Course (Technical courses for experimental classes)
Study Period and Credits: 64 hours / 4 credits

课程介绍
由建筑信息模型（BIM）方法（Revit 教学）、建筑参数化建模方法（Grasshopper 教学）和建筑结构概念设计方法（建筑学与结构工程交叉领域教学）三个模块组成。BIM 模块强化了信息模型如何提高建筑设计效率与质量的教学，巩固了建筑构造的理解；建筑参数化建模模块强化了基于建筑设计逻辑的方法拆解与组合；建筑结构概念设计模块加强了结构理论知识快速融入建筑设计的方法教学，初步形成结构化思维方式。

教学目标
BIM 模块为学生在建筑数字孪生前沿探索提供准备；建筑参数化建模模块为学生在建筑设计与计算机算法的交叉领域探索提供准备；建筑结构概念设计模块为学生在建筑学与结构工程的可持续整合探索提供准备。

Course description
The course is composed of three modules: Building Information Modeling (BIM) method (Revit teaching), architectural parametric modeling method (Grasshopper teaching), and the conceptual design method for building structures (teaching in the interdisciplinary field of architecture and structural engineering). The BIM module strengthens the teaching of how information models enhance the efficiency and quality of architectural design, consolidating the understanding of architectural construction; the architectural parametric modeling module emphasizes the breakdown and combination of methods based on architectural design logic; the conceptual design module for building structures enhances the teaching method of quickly integrating structural theory knowledge into architectural design, initially forming a structured way of thinking.

Teaching objectives
The BIM module prepares students for exploration at the forefront of architectural digital twins; the architectural parametric modeling module prepares students for exploration in the interdisciplinary field of architectural design and computer algorithms; the conceptual design module for building structures prepares students for exploration in the sustainable integration of architecture and structural engineering.

本科二年级 / 本科三年级
建筑物理 / 建筑技术（二）：声光热 · 吴蔚
课程类型：必修
学时 / 学分：32 学时 / 2 学分

Undergraduate Program 2nd Year/ 3rd Year
BUILDING PHYSICS / ARCHITECTURAL TECHNOLOGY 2: SOUND, LIGHT AND HEAT
WU Wei
Type: Required Course
Study Period and Credits: 32 hours / 2 credits

课程介绍
本课程是针对南京大学建筑学本科学生来设计的。课程介绍了建筑热工学、建筑光学、建筑声学中的基本概念和基本原理，帮助学生掌握建筑的热环境、声环境、光环境的质量评价方法，以及相关的国家标准。在了解建筑节能设计的基本原则和理论后，学生应具备在计算机模拟技术的帮助下，完成一定建筑节能设计的能力。

Course description
This course is designed for the undergraduates at the School of Architecture, Nanjing University. The course introduces the basic concepts and principles of architectural thermal engineering, architectural optics, and architectural acoustics, helping students master the quality evaluation methods of the thermal environment, acoustic environment, and light environment of buildings, as well as the relevant national standards. After understanding the basic principles and theories of building energy efficiency design, students should have the ability to complete certain building energy efficiency design with the help of computer simulation technology.

本科二年级 / 本科三年级
建筑设备 / 建筑技术（三）：水电暖・吴蔚
课程类型：必修
学时 / 学分：32 学时 / 2 学分

Undergraduate Program 2nd Year/ 3rd Year
CONSTRUCTION EQUIPMENT/ ARCHITECTURAL TECHNOLOGY 3: WATER, ELECTRICITY AND HEATING • WU Wei
Type: Required Course
Study Period and Credits: 32 hours / 2 credits

课程介绍
本课程是针对南京大学建筑学院本科学生来设计的。课程介绍了建筑给水排水系统、采暖通风与空气调节系统、电气工程的基本理论、基本知识和基本技能，使学生掌握熟练阅读水电、暖工程图的能力，熟悉水电及消防的设计、施工规范，了解燃气供应、安全用电及建筑防火、防雷的初步知识。

Course description
This course is designed for the undergraduates at the School of Architecture, Nanjing University. The course introduces the basic theories, knowledge, and skills of building water supply and drainage systems, heating, ventilation, and air conditioning systems, and electrical engineering. It aims to enable students to master the ability to proficiently read water, electricity, and HVAC engineering drawings, become familiar with the design and construction specifications of water, electricity, and fire protection, and understand the preliminary knowledge of gas supply, safe electricity use, building fire prevention, and lightning protection.

本科三年级
建筑技术（一）：建构设计・傅筱
课程类型：必修
学时 / 学分：36 学时 / 2 学分

Undergraduate Program 3rd Year
ARCHITECTURAL TECHNOLOGY 1: CONSTRUCTION DESIGN • FU Xiao
Type: Required Course
Study Period and Credits:36 hours / 2 credits

课程介绍
本课程是建筑学专业本科生的专业主干课程。本课程的任务主要是以建筑师的工作性质为基础，讨论一个建筑生成过程中最基本的三大技术支撑（结构、构造、施工）的原理性知识要点，以及它们在建筑实践中的相互关系。

Course description
The course is a major course for the undergraduates of architecture. The main purpose of this course is based on the nature of the architect's work, to discuss the principle knowledge points of the basic three technical supports in the process of generating construction (structure, construction, execution), and their mutual relations in the architectural practice.

本科三年级 / 本科四年级
建成环境科学・施珊珊
课程类型：选修
学时 / 学分：32 学时 / 2 学分

Undergraduate Program 3rd Year/ 4th Year
BUILT ENVIRONMENT SCIENCE • SHI Shanshan
Type: Elective Course
Study Period and Credits: 32 hours / 2 credits

课程介绍
建成环境领域科学研究能力的培养：帮助本科生建立科学研究的基本概念，了解建成环境领域的前沿发展方向，掌握基本研究方法在建成环境领域的应用方式，最终初步具备设计、执行、分析建成环境领域科学研究的能力。建成环境科学领域研究成果驱动建筑设计能力的培养；系统介绍建筑物理、建筑能源及建筑环境等前沿发展方向研究成果，搭建研究成果与设计驱动的关联关系，培养学生基于建成环境科学领域前沿研究成果回应设计实际问题的能力。

Course description
The cultivation of scientific research capabilities in the field of built environment: To help the undergraduates establish basic concepts of scientific research, understand the cutting-edge development directions in the field of built environment, master the application of basic research methods in the field of built environment, and ultimately possess the ability to design, execute, and analyze scientific research in the field of built environment. The cultivation of architectural design ability driven by research achievements in the field of built environment science: Systematically introduce research achievements in cutting-edge development directions such as building physics, architectural energy, and architectural environment, establish a correlation between research achievements and design driven, and cultivate students' ability to respond to practical design problems based on cutting-edge research achievements in the field of built environment science.

本科四年级
CAAD 理论与实践（二）・吉国华
课程类型：选修
学时 / 学分：18 学时 / 1 学分

Undergraduate Program 4th Year
THEORY AND PRACTICE OF CAAD 2 • JI Guohua
Type: Elective Course
Study Period and Credits: 18 hours / 1 credits

教学目标
随着计算机辅助建筑设计技术的快速发展，当前数字技术在建筑设计中的角色逐渐从辅助绘图转向了真正的辅助设计，并引发了设计的革命和建筑的形式创新。本课程讲授 Grasshopper 参数化编程建模方法以及相关的几何知识，让学生在掌握参数化编程建模技术的同时，增强以理性的过程思维方式分析和解决设计问题的能力，为数字建筑设计和数字建造打下必要的基础。

课程内容
课程讲授基于 Rhinoceros 的算法编程平台 Grasshopper 的参数化建模方法，包括各类运算器的功能与使用、图形的生成与分析、数据的结构与组织、各类建模的思路与方法，以及相应的数学与计算机编程知识。

Teaching objectives
With the rapid development of computer-aided architectural design technology, the role of digital technology in architectural design has gradually shifted from auxiliary drawing to truly auxiliary design, and has triggered a revolution in design and innovation in architectural form. This course teaches Grasshopper parametric programming modeling methods and related geometric knowledge, enabling students to master parametric programming modeling techniques while enhancing their ability to analyze and solve design problems in a rational process thinking manner, laying a necessary foundation for digital building design and digital construction.

Course content
The course teaches the parameterized modeling methods of Grasshopper, an algorithm programming platform based on Rhinoceros, including the functions and usage of various algorithms, graph generation and analysis, data structure and organization, various modeling ideas and methods, as well as corresponding mathematical and computer programming knowledge.

本科四年级
设计研究与环境行为 · 窦平平
课程类型：选修
学时 / 学分：32 学时 / 2 学分

Undergraduate Program 4th Year
DESIGN RESEARCH AND ENVIRONMENTAL BEHAVIOR · DOU Pingping
Type: Elective Course
Study Period and Credits: 32 hours / 2 credits

课程介绍
课程密切结合当代技术应用与城市发展前景，面向未来人居的新挑战，开展前沿性的理论思考与辨析；突破学科界限，立足多学科交叉协同，以学科核心问题带动跨学科思维和跨界应用，帮助学生进行创新思考；将历史与理论作为思维工具，以设计研究为方法，重点培养应用研究能力，强调教学的实验性，引导学生通过独立思考对建筑形成更加个人化和批判性的认识，建立设计研究的工作习惯。

Course description
The course closely integrates contemporary technology applications and urban development prospects, faces new challenges in future human settlements, and conducts cutting-edge theoretical thinking and analysis, breaking through disciplinary boundaries, based on interdisciplinary collaboration, using core disciplinary issues to drive interdisciplinary thinking and applications, and helping students engage in innovative thinking; using history and theory as thinking tools, using design research as a method, focusing on cultivating applied research abilities, emphasizing the experimental nature of teaching, guiding students to form a more personalized and critical understanding of architecture through independent thinking, and establishing work habits for design research.

本科四年级
建设工程项目管理 · 谢明瑞
课程类型：选修
学时 / 学分：36 学时 / 2 学分

Undergraduate Program 4th Year
CONSTRUCTION PROJECT MANAGEMENT · XIE Mingrui
Type: Elective Course
Study Period and Credits: 36 hours / 2 credits

课程介绍
帮助学生系统掌握建设工程项目管理的基本概念、理论体系和管理方法，了解建筑规划设计在建设工程项目中的地位、特点和重要性。拓展建筑学专业学生的基本知识结构和学生的发展方向。

Course description
To help students systematically master the basic concept, theoretical system and management method of construction project management, understand the position, characteristics and importance of architectural planning design in the construction engineering project.
To extend the basic knowledge structure and the development direction of students majoring in architecture.

研究生一年级
传热学与计算流体力学基础 · 郜志
课程类型：选修
学时 / 学分：36 学时 / 2 学分

Postgraduate Program 1st Year
FUNDAMENTALS OF HEAT TRANSFER AND COMPUTATIONAL FLUID DYNAMICS · GAO Zhi
Type: Elective Course
Study Period and Credits: 36 hours / 2 credits

课程介绍
本课程的主要任务是使建筑学 / 建筑技术学专业的学生掌握传热学和计算流体力学的基本概念和基础知识，通过课程教学，使学生熟悉传热学中导热、对流和辐射的经典理论，并了解传热学与计算流体力学的实际应用和最新研究进展，为建筑能源和环境系统的计算和模拟打下坚实的理论基础。教学尽量简化传热学和计算流体力学经典课程中复杂公式的推导过程，而着重于如何解决建筑能源与建筑环境中涉及流体流动和传热的实际应用问题。

Course description
The tasks of this course are to enable students majoring in building science and engineering / architectural technology to proficiently master the basic concept and knowledge of heat transfer and computational fluid dynamics (CFD). Through course teaching, students will become familiar with the classical theories of conduction, convection and radiation in heat transfer, and understand the practical application and latest research progress of heat transfer and CFD. This teaching aims to simplify the derivation process of complex formulas in classic courses of heat transfer and computational fluid dynamics, with a focus on how to solve practical application problems involving fluid flow and heat transfer in building energy and building environment.

研究生一年级
材料与建造 · 冯金龙
课程类型：选修
学时 / 学分：18 学时 / 1 学分

Postgraduate Program 1st Year
MATERIALS AND CONSTRUCTION · FENG Jinlong
Type: Elective Course
Study Period and Credits: 18 hours / 1 credit

课程介绍
本课程介绍现代建筑技术的发展过程，论述现代建筑技术及其美学观念对建筑设计的重要作用，探讨由材料、结构和构造方式所形成的建筑建造的逻辑方式，研究建筑形式产生的物质技术基础，诠释现代建筑的建构理论与研究方法。

Course description
This course introduces the development process of modern architectural technology and discusses the important role played by the modern architectural technology and its aesthetic concept in the architectural design, explores the logical methods of construction of the architecture formed by materials, structure and construction, and studies the material and technical basis for the creation of architectural form, interpreting the construction theory and research method for modern architecture.

Postgraduate Program 1st Year
COMPUTER-AIDED TECHNOLOGY · JI Guohua
Type: Required Course
Study Period and Credits: 18 hours / 1 credit

Course description
With the rapid development of computer-aided architectural design technology, the role of digital technology in architectural design has gradually shifted from auxiliary drawing to true auxiliary design, triggering a revolution in design and innovation in architectural form. This course introduces methods of Grasshopper parametric programming and modeling and relevant geometric knowledge, allows students enhance the ability to analyze and solve designing problems with a rational thinking way while mastering parametric programming modeling technology, and build necessary foundation for digital building design and digital construction.
This course teaches parametric modeling methods based on Grasshopper, a algorithmic programming platform for Rhinoceros, including functions and application of all kinds of arithmetic units, pattern formation and analysis, structure and organization of data, various thoughts and methods of modeling, and related knowledge of mathematics and computer programming.

Postgraduate Program 1st Year
GREEN BUILDING TECHNOLOGY · LIAGN Weihui
Type: Elective Course
Study Period and Credits: 36 hours / 2 credits

Course description
The primary objective of this course is to give students a comprehensive understanding of green building design. Through the course instruction, students will gain insight into the concepts and aspect associated with green buildings, understand the fundamental principles of green building design, and acquire knowledge of advanced green building technologies and energy-efficient construction practices. This course aims to help students recognize that green buildings must be adapted to local climates and resources, develop the ability to apply green building technology with critical thinking, and provide a solid foundation for future research and professional endeavors in the field of green buildings. The main content and chapters of the course include an overview of green buildings, the relationship between buildings and climate, indoor and outdoor environmental analysis and design techniques (including heat, wind, and natural ventilation of buildings, sunshine and lighting environment in building), energy-saving design and technology, the utilization of renewable energy and water resources and green buildings, and green evaluation standards and cases.

认知实习
COGNITIVE INTERNSHIP

本科三年级
乡村振兴建设实践・梁宇舒
课程类型：选修
学时 / 学分：32 学时 / 2 学分

Undergraduate Program 3rd Year
PRACTICE OF RURAL REVITALIZATION CONSTRUCTION • LIANG Yushu
Type: Elective Course
Study Period and Credits: 32 hours / 2 credits

教学目标
　　本课程的任务主要是通过乡村现场调研以及驻场设计实践，培养学生借助多学科手段分析和认知实际社会问题的能力，并且综合运用建筑设计、乡村规划等方法，针对实际场景提出可行的解决方案，加深图纸设计与实际建造、社会服务之间的关联性理解。

课程内容
　　1. 行前培训讲座；
　　2. 乡村调研及民居考察；
　　3. 乡村调研报告讨论；
　　4. 乡村振兴规划与建筑设计；
　　5. 乡村振兴工作营成果汇报。

Teaching objectives
The main task of this course is to cultivate students' ability to analyze and recognize practical social problems through multi-disciplinary methods through rural on-site research and on-site design practice, and to comprehensively apply methods such as architectural design and rural planning to propose feasible solutions for practical scenarios, deepening the understanding of the correlation between drawing design, actual construction, and social services.

Course content
1. Pre-departure training lectures;
2. Rural research and residential investigation;
3. Discussion on rural research reports;
4. Rural revitalization planning and architectural design;
5. Report on the achievements of the rural revitalization work camp.

本科三年级
城乡认知实习・冯建喜　王洁琼
课程类型：选修
学时 / 学分：32 学时 / 2 学分

Undergraduate Program 3rd Year
URBAN AND RURAL COGNITIVE INTERNSHIP
• FENG Jianxi, WANG Jieqiong
Type: Elective Course
Study Period and Credits: 32 hours / 2 credits

教学目标
　　1. 系统学习区域、城市与建筑认知的理念与方法，同时深入了解建筑设计与评价的基本方法和原则。
　　2. 了解典型城市区域关系、发展脉络、用地布局与空间结构、城市优秀地段和建筑空间的尺度和形式。
　　3. 学习成功的区域规划、城市和建筑设计的范例，也认知失败的区域关系、城市空间和建筑，从中汲取经验教训，增加感性认识，开阔眼界。
　　4. 掌握评价现实城市与建筑的方法，独立完成城市和建筑空间的认知分析报告及学术论文。
　　5. 熟悉具体区域规划、城市规划和建筑设计过程中解决实际问题的方法，同时为下一阶段的城市规划专业学习打下良好的基础。

Teaching objectives
1. Systematically study the concepts and methods of regional, urban, and architectural cognition, while delving into the basic methods and principles of architectural design and evaluation.
2. Understand typical urban regional relationships, development context, land layout and spatial structure, as well as the scale and form of excellent urban areas and architectural spaces.
3. Learn examples of successful regional planning, urban and architectural design, as well as recognize failed regional relationships, urban spaces and architecture, draw lessons from them, increase emotional understanding, and broaden horizons.
4. Master the methods of evaluating real cities and buildings, independently complete cognitive analysis reports and academic papers on urban and architectural spaces.
5. Be familiar with the methods of solving practical problems in specific regional planning, urban planning, and architectural design processes, while laying a solid foundation for the next stage of urban planning professional learning.

本科三年级
城市更新规划・陈浩
课程类型：选修
学时 / 学分：32 学时 / 2 学分

Undergraduate Program 3rd Year
URBAN REGENERATION PLANNING
• CHEN Hao
Type: Elective Course
Study Period and Credits: 32 hours / 2 credits

课程介绍
　　本课程的学习任务主要分为四大模块：基础理论知识、基础调研分析、经典案例分析、规划设计实践。基础理论知识模块以讲授基本理论、技术方法和学科行业动态为主，在保证相关基础知识得到全面介绍的同时增强教学的前沿性和时效性；基础调查分析模块以讲授城市更新基础调查内容与方法为基础，为学生提供真实的传统居住、工业、公共空间三类实践场地，在教师引导下发挥学生的创造性，形成对于设计场地及更新任务的深刻认知；经典案例分析模块以邀请学界和行业专家介绍经典案例的成败经验为主，加深学生对于前沿理念和实践的理解；规划设计实践模块在调查分析的基础上对传统居住、工业和综合功能街区进行更新规划设计与实施方案创作，并邀请专家对学生课程成果做点评和指导。

Course description
The learning tasks of this course are mainly divided into four modules: basic theoretical knowledge, basic research and analysis, classic case study, and planning and design practice. The basic theoretical knowledge module focuses on teaching basic theories, technical methods and industry dynamics, which ensures that the relevant basic knowledge is comprehensively introduced while enhancing the cutting-edge and timeliness of teaching; the basic investigation and analysis module is based on the teaching of the content and methodology of basic urban renewal surveys, which provides students with three types of real practice sites, namely, traditional residential, industrial, and public spaces, and allows students to utilize their creativity and develop a deep understanding of the design sites and renewal tasks; the classic case study module focuses on introducing the success or failure experience of the classic cases by academic and industry experts, so as to deepen the students' understanding of the cutting-edge concepts and practices; the planning and designing practice module carries out the renewal planning, designing and implementation of the traditional residential, industrial and comprehensive functional blocks on the basis of investigation and analysis, and invites experts to comment and guide the students' achievements in the course.

本科四年级
古建筑测绘 · 赵辰 史文娟 赵潇欣
课程类型：选修
学时 / 学分：36 学时 / 2 学分

Undergraduate Program 4th Year
SURVEYING AND MAPPING OF ANCIENT BUILDINGS · ZHAO Chen, SHI Wenjuan, ZHAO Xiaoxin
Type: Elective Course
Study Period and Credits: 36 hours / 2 credits

教学目标
　　本课程是建筑学专业本科生的专业基础理论课程。其任务是使学生切实理解中国传统建筑结构体系、构造关系及比例尺度等基本概念，培养学生对传统建筑年代鉴定和价值判断的基本技能。

教学内容
　　通过室外作业和室内工作两个阶段，完成现场测绘和整理图纸报告两个环节："测"，观测量取现场实物的尺寸数据；"绘"，根据测量数据与草图整理绘制完备的测绘图纸，最终完成个人独立的答辩报告。
　　第一阶段：抵达现场，听讲座（第1天）。
　　第二阶段：现场工作，各小组分工协作（第2—7天）。
　　第三阶段：图纸绘制，小组资料归档、排版（第8—14天）。
　　第四阶段：答辩准备，打印图纸、个人独立研究（第15—20天）。
　　第五阶段：交图答辩（第21天）。

Teaching objectives
This course is a professional basic theory course for undergraduates majoring in architecture. The mission of this course is to enable students to truly understand the basic concepts such as the building structural system, structural relationship, and proportional scale of traditional Chinese buildings, and to cultivate students' basic skills in age identification and value judgment of traditional buildings.

Course content
Through the two stages of outdoor work and indoor work, the two links of on-site surveying and mapping and drawing report preparation are completed: "Measuring", observing and measuring the dimensional data of the actual objects on site; "Drawing", organizing and drawing complete surveying and mapping drawings based on the measurement data and sketches, and finally complete a personal and independent defense report.
Phase 1: Arriving at site and attending lectures (Day 1).
Phase 2: On-site work, division of labor and collaboration among each group (Days 2 to 7).
Phase 3: Drawing drawings, archiving and typesetting of group materials (Days 8 to 14).
Phase 4: Defense preparation, printing drawings and independent research (Days 15–20).
Phase 5: Submission defense (Day 21).

本科四年级
工地实习 · 傅筱 吴佳维
课程类型：选修
学时 / 学分：36 学时 / 2 学分

Undergraduate Program 4th Year
CONSTRUCTION SITE INTERNSHIP · FU Xiao, WU Jiawei
Type: Elective Course
Study Period and Credits: 36 hours / 2 credits

教学目标
　　本课程实习的训练目的是加深学生对建筑从图纸到实际建造过程的认识和理解，重点理解图纸与建造的关联。本课程的任务主要是通过工地现场考察，让学生了解建筑材料的生产过程，对建筑实际建造流程有一定的直观认识；让学生实地观察某一建筑构件的建造过程，绘制建筑构件施工图纸；让学生掌握建筑面层材料与建筑主体结合的构造关系，理解建筑材料运用与设计表达之间的关联性。

教学内容
　　课堂讲授：由教师课堂讲授相关的技术知识。
　　厂家调研：教师和助教带领学生在建材厂进行考察和学习。
　　知识梳理：结合调研收集到的资料和教师的讲解，梳理从原材料到建材生产、现场安装的全过程。
　　构造轴测图绘制：进一步搜索图集、案例，绘制相关节点的构造轴测图。
　　实习报告：根据厂家考察和课堂讲授的学习，完成实习报告。

Teaching objectives
The training purpose of on the construction site internship is to deepen students' knowledge and understanding of the process from architectural drawings to actual construction with a focus on understanding the connection between drawings and construction. The main tasks of this course are as follows: Let students understand the production process of building materials and have a certain intuitive understanding of the actual construction process of buildings through on-site inspection; let students observe the construction process of a specific building component on the spot and draw construction drawings of the components; let students master the structural relationship between surface materials and the main body of the building, and understand the correlation between the use of building materials and design expression.

Teaching content
Classroom lectures: Teachers will give lectures on relevant technical knowledge in class.
Factory research: Teachers and teaching assistants will lead students to conduct inspections and study in building materials factories.
Knowledge combing: Comb the whole process from raw materials to building material production and on-site installation in combination with the information collected in the research and the teacher's explanation.
Structural axonometric drawing: Further search for atlases and cases, and draw structural axonometric drawings of relevant nodes.
Internship report: According to the manufacturer's inspection and classroom lectures, complete the internship report.

其他
MISCELLANEA

讲座
Lectures

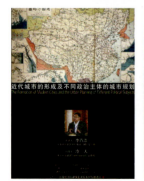

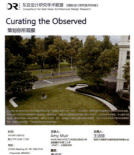

硕士学位论文列表
List of Master's Theses

研究生姓名	研究生论文标题	导师姓名
刘亲贤	城市历史片区中建筑改造设计研究——以小西湖7号单元南二区建筑改造为例	丁沃沃
宋贻泽	城市更新背景下快速路空间综合利用设计研究	丁沃沃
徐佳楠	固废夯制建筑构造设计研究——以陈墩村史馆设计为例	丁沃沃
于文爽	基于形状语法的相交空间生成规则研究	丁沃沃
陈 颖	基于密度评估的紧凑城市形态研究	丁沃沃
杨瑞侃	改革开放以来境外建筑师及其在中国境内的设计作品对中国建筑创作的影响研究	丁沃沃
费元丽	基于亲生物策略的疗愈环境循证设计研究——图像智能识别作为一种新的途径	窦平平
王明珠	声景建筑的多维感知测度及其可视化与空间化表达	窦平平
于瀚清	可穿戴交互技术下的城市视听环境情绪感知与可视化表达	窦平平
蒋 哲	城市更新中的文化宜复双驱动设计策略——声景营造作为一种途径	窦平平
潘 晴	南京大学鼓楼校区南园配套服务设施设计实践——钢结构构造节点设计及建造过程	冯金龙
唐 敏	南京大学鼓楼校区计算机中心改造设计实践	冯金龙
谢文俊	苏州光福镇九年一贯制学校设计——地下接送空间设计研究	冯金龙
赵亚迪	南京市紫核龙王山小学设计——基于教育变革背景下小学非正式学习空间设计研究	冯金龙
周理洁	南京大学鼓楼校区图书馆更新改造之设计与实践	冯金龙
冯 智	宿迁古渠遗址博物馆建筑设计——遗址博物馆展览空间设计研究	冯金龙、冯正功
王瑞蓬	大运河沿岸工业遗存活化改造的设计策略研究——宿迁运河湾精品酒店设计	冯金龙、冯正功
崔晓伟	基于过渡空间形态的公共建筑被动式设计策略研究——以南京中医药大学会议中心为例	冯金龙、谢明瑞
刘雨田	南京溧水薛李东路幼儿园项目设计——夏热冬冷地区幼儿园北廊空间引入阳光的设计研究	傅 筱
孙 杰	复合化表演中心设计研究——溧水城南小学表演中心设计	傅 筱
翁鸿祎	"双减"背景下的小学图艺中心设计研究——以溧水城南小学为例	傅 筱
朱凌云	中小学体育馆自然通风优化设计研究——以南京溧水城南小学体育馆为例	傅 筱
王 路	周边建筑遮挡条件下居住区通风条件的风热耦合数值模拟边界设定方法研究	傅 筱
陈予婧	城市小学地下接送系统送学等候流线量化设计研究——以南京溧水城南小学为例	傅 筱
胡永裕	夏热冬冷地区办公建筑窗户应用半透明辐射制冷薄膜对室内热环境及能耗的影响研究	郜 志
李 倩	超越罗马——米开朗琪罗在罗马的教堂建筑设计研究	胡 恒
曾敬淇	基于冷岛效应的高密度城市肌理形态优化设计原理与策略研究	胡友培
吕广彤	基于结构找形优化的户外剧场设计研究——以响堂村水边剧场项目为例	胡友培
王 琪	荷兰结构主义聚落型建筑设计方法研究	胡友培
朱激清	从"高层城市"到"新城市"——路德维希·希尔伯塞默的城市概念研究	胡友培
陈宇帆	结合低碳策略的夏热冬冷地区乡村建筑更新设计研究——马鞍山市博望区横山林场建筑改造更新设计	华晓宁
丁嘉欣	后疫情时代基于健康城市理念的社区体育基础设施设计研究——珠江路安康苑居住区规划与单体设计	华晓宁
刘奕孜	城市基础设施场所形成机制研究——以南京为例	华晓宁
秦钢强	基于Stable Diffusion的住宅平面生成研究	吉国华
丘雨辰	基于多面体的三维图解静力学找形研究暨南京大学南园景观亭设计	吉国华
刘雪寒	基于太阳辐射和Pix2Pix模型的办公建筑采光性能预测方法研究	吉国华
邵 桐	建筑形体的非正交化及在性能化建筑设计中的应用	吉国华
杨东来	基于树图的建筑与地块平面生成方法研究	吉国华
仇佳豪	关于道胜堂旧址早期规划及代表建筑的历史研究	冷 天

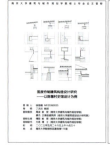

研究生姓名	研究生论文标题	导师姓名
张塑琪	基于使用者需求的高校既有公共教学空间智慧化改造策略研究——以南京大学新教楼为例	冷　天
黄翊婕	近代南京银行建筑营建机制研究——以上海商业储蓄银行南京分行为例	冷　天
雷　畅	近代水泥鱼鳞瓦屋面体系的营造工艺及技术特征初探	冷　天
谢　菲	近代"泰山面砖"之产品、生产及应用研究	冷　天
许佑龄（许龄）	高校宿舍室内空气与典型建材气味主客观评价和方法优化研究	梁卫辉
白珂嘉	高校体育馆暖通空调设备空间整合设计研究——以南京邮电大学仙林校区新体育馆优化设计为例	刘　铨、冯金龙
马丹艺	位置媒介应用对城市漫游体验影响的实证研究——以南京文学小路为例	鲁安东
王雨嘉	多层次声信息对城市公园文化场所的场所感影响研究——以南京文学小路所节点为例	鲁安东
张梦冉	"双碳"背景下展览空间的快速建造研究——以第九届城市/建筑深港双城双年展为例	鲁安东
龚泰冉	苏州古城37、38街坊宗教空间变迁关系研究	鲁安东
庞馨怡	紫金山玄武湖轴线体系的演变及规律研究	鲁安东
王　蕾	玄武湖文化景观的演进与构成研究	鲁安东
马致远	当代城市中心区步行公共空间形态特征表述方法研究——以南京新街口为例	唐　莲、丁沃沃
罗紫娟	低碳导向下的西方城市街道更新案例分析	童滋雨
朱孟阳	基于CIM的居住类街区更新设计辅助系统的构建研究——以江西省瑞昌市城隍庙片区为例	童滋雨
吴子豪	基于局地气候分区的城市形态演变模式及其地表热环境效应研究——以南京市为例	童滋雨
张云松	基于网格的拓扑互锁结构生成方法研究	童滋雨
盛泽明	迪恩·霍克斯的建筑环境思想研究	王骏阳
朱辰浩	建构表现之"轻"：隈研吾建筑设计作品研究	王骏阳
林之茜	五台山佛光寺场地营建研究	王骏阳
陈铭行	可转换为居住空间的多层独栋小型商业空间设计研究——以苏浙皖商贸中心茶文化工坊为例	张　雷
赖泽贤	溧阳苏浙皖农产品商贸中心设计——高层商业建筑空间向居住空间转换的灵活设计研究	张　雷
杨淑钏	场所记忆视角下老旧街区改造设计研究——以南京军人俱乐部改造项目为例	张　雷
余沁蔓	城市中心区旧建筑更新中的立体化公共空间设计研究——以南京市新街口艺术大楼改造项目为例	张　雷
袁振香	现代粮仓园区体验性设计研究——以南京江宁凤凰山粮仓园区设计为例	张　雷
赵济宁	石家庄太平河城市片区展示中心设计——悬浮式城市公共空间研究	张　雷
龚豪辉	南京利济巷公共文化空间项目历史沿革与新条件下的再设计	赵　辰
李　昂	江阴周庄镇中心地块低层高密传统尺度住宅区规划设计研究	赵　辰
任钰佳	工业竹轻型框架体系建筑的应用拓展设计研究	赵　辰
邢雨辰	"空间家具"——基于工业化竹材应用的集成结构设计研究	赵　辰
杨　岚	中国传统工业遗产地遗产价值体系初探——以禹州钧瓷工业遗产地为例	赵　辰
陈婧秋	"堂横屋"改扩建之形态与建构研究——以龙岩云山村村民自主更新案例为对象	赵　辰
程　慧	基于形态类型学的盐城市收成村肌理修复与建筑更新策略	周　凌
方奕璇	文脉传承视角下阜宁单家港村落空间更新设计研究	周　凌
于智超	连云港西连岛酒店组团设计及其自然通风性能分析研究	周　凌
赵子文	公共服务视角下乡村民宿共享设计——以营盘圩乡村民宿为例	周　凌
李晓云	黄河故道沿线乡村聚落空寂形态研究——以阜宁县为例	周　凌
赵琳芝	黔中地区传统城镇街区空间形态与微气候环境的模拟研究——以荔波老城为例	周　凌
周宇飞	生活方式变迁视野下的近代上海公寓住宅研究	周　凌、黄华青

在校学生名单
List of Students

本科生 Undergraduates

2019级学生 / Students 2019

高禾雨 GAO Heyu	石珂千 SHI Keqian	周昌赫 ZHOU Changhe
高赵龙 GAO Zhaolong	唐诗诗 TANG Shishi	上原舜平 SHANGYUAN Shunping
顾 靓 GU Liang	王思戎 WANG Sirong	麦吾兰江·穆合塔尔 Maiwulanjiang MUHETAR
黄辰逸 HUANG Chenyi	王智坚 WANG Zhijian	
黄小东 HUANG Xiaodong	王梓蔚 WANG Ziwei	
黄煜东 HUANG Yudong	袁 泽 YUAN Ze	
邱雨婷 QIU Yuting	张楚杭 ZHANG Chuhang	

2020级学生 / Students 2020

陈浏毓 CHEN Liuyu	华羽纶 HUA Yuguan	刘卓然 LIU Zhuoran	吴嘉文 WU Jiawen
陈沈婷 CHEN Shenting	黄淑睿 HUAGN Shurui	陆星宇 LU Xingyu	杨曦睿 YANG Xirui
陈璇霖 CHEN Xuanlin	李静怡 LI Jingyi	钱梦南 QIAN Mengnan	袁欣鹏 YUAN Xinpeng
陈 琤 CHEN Cheng	李沛熹 LI Peixi	孙昊天 SUN Haotian	张嘉木 ZHANG Jiamu
高 晴 GAO Qing	李若松 LI Ruosong	沈至文 SHEN Zhiwen	张伊儿 ZHANG Yi'er
顾 林 GU Lin	刘珩歆 LIU Hengxin	王天赐 WANG Tianci	
何德林 HE Delin	刘晓斌 LIU Xiaobin	王天歌 WANG Tian'ge	

2021级学生 / Students 2021

阿 旦 A Dan	邹子午 ZOU Ziwu	黄文郡 HUANG Wenjun	李雨芊 LI Yuqian	汪祉健 WANG Zhijian	杨雯文 YANG Wenwen	周皓誉 ZHOU Haoyu
陈雅琪 CHEN Yaqi	步欣洁 BU Xinjie	江炫烨 JIANG Xuanye	刘新桐 LIU Xintong	王翔宇 WANG Xiangyu	杨子江 YANG Zijiang	周锦润 ZHOU Jinrun
黄静雯 HUANG Jingwen	陈 郝 CHEN He	金 毅 JIN Yi	刘子宣 LIU Zixuan	王艺楠 WANG Yinan	张纪闻 ZHANG Jiwen	周抒秋 ZHOU Shuqiu
黎宇航 LI Yuhang	陈 可 CHEN Ke	孔令璇 KONG Lingxuan	阮锦涛 RUAN Jintao	王雨晴 WANG Yuqing	张芮宁 ZHANG Ruining	朱 轶 ZHU Yi
彭 鑫 PENG Xin	陈宇轩 CHEN Yuxuan	李俊杰 LI Junjie	佘柯宇 SHE Keyu	奚 凡 XI Fan	张昱程 ZHANG Yucheng	
申 翱 SHEN Ao	何若然 HE Ruoran	李梦麟 LI Menglin	万飞扬 WAN Feiyang	夏知愚 XIA Zhiyu	赵 同 ZHAO Tong	
唐扬航 TANG Yanghang	黄思宇 HUANG Siyu	李宇博 LI Yubo	万家良 WANG Jialiang	颜星林 YAN Xinglin	郑可薇 ZHENG Kewei	

2022级学生 / Students 2022

陈彦博 CHEN Yanbo	茆溢轩 MAO Yixuan	张效儒 ZHANG Xiaoru	云旦加措 YUNDAN Jiacuo
江玉洁 JIANG Yujie	陶新欣 TAO Xinxin	张耀天 ZHANG Yaotian	扎西央宗 TASHI Yangzong
李伟康 LI Weikang	王明宇 WANG Mingyu	赵志勇 ZHAO Zhiyong	
凌子昂 LING Zi'ang	王馨柠 WANG Xinning	郑海培 ZHENG Haipei	
刘超然 LIU Chaoran	王奕婷 WANG Yiting	朱奕帆 ZHU Yifan	
刘 澈 LIU Che	王玉欣 WANG Yuxin	次旦央宗 TAETAN Yangzong	
刘至文 LIU Zhiwen	吴思婷 WU Siting	红得孜·我曼 Hongdezi WOMAN	

研究生 Postgraduates

白珂嘉 BAI Kejia	丁嘉欣 DING Jiaxin	蒋哲 JIANG Zhe	刘奕孜 LIU Yizi	仇佳豪 QIU Jiahao	唐敏 TANG Min	谢文俊 XIE Wenjun	余沁蔓 YU Qinman	赵琳芝 ZHAO Linzhi
陈靖秋 CHEN Jingqiu	丁明昊 DING Minghao	赖泽贤 LAI Zexian	刘雨田 LIU Yutian	丘雨辰 QIU Yuchen	王明珠 WANG Mingzhu	徐福锁 XU Fusuo	于文爽 YU Wenshuang	赵亚迪 ZHAO Yadi
陈铭行 CHEN Mingxing	方奕璇 FANG Yixuan	雷畅 LEI Chang	罗紫娟 LUO Zijuan	任钰佳 REN Yujia	王琪 WANG Qi	许龄 XU Ling	于智超 YU Zhichao	赵子文 ZHAO Ziwen
陈茜 CHEN Qian	费元丽 FEI Yuanli	李昂 LI Ang	吕广彤 LU Guangtong	邵桐 SHAO Tong	王瑞蓬 WANG Ruipeng	徐佳楠 XU Jianan	袁振香 YUAN Zhenxiang	周理洁 ZHOU Lijie
陈颖 CHEN Ying	冯智 FENG Zhi	李倩 LI Qian	马丹艺 MA Danyi	盛泽明 SHENG Zeming	王雨嘉 WANG Yujia	杨东来 YANG Donglai	曾敬淇 ZENG Jingqi	周宇飞 ZHOU Yufei
陈予靖 CHEN Yujing	龚豪辉 GONG Haohui	李晓云 LI Xiaoyun	马致远 MA Zhiyuan	宋贻泽 SONG Yize	翁鸿祎 WENG Hongyi	杨岚 YANG Lan	张梦冉 ZHANG Mengran	朱辰浩 ZHU Chenhao
陈宇帆 CHEN Yufan	龚泰冉 GONG Tairan	林之茜 LIN Zhiqian	潘晴 PAN Qing	孙杰 SUN Jie	吴子豪 WU Zihao	杨瑞侃 YANG Ruikan	张塑琪 ZHANG Suqi	朱激清 ZHU Jiqing
程慧 CHENG Hui	胡永裕 HU Yongyu	刘沁贤 LIU Qinxian	庞馨怡 PANG Xinyi	王亲贤 WANG Lei	谢非 XIE Fei	杨淑钏 YANG Shuchuan	张云松 ZHANG Yunsong	朱凌云 ZHU Lingyun
崔晓伟 CUI Xiaowei	黄翔婕 HUANG Yijie	刘雪寒 LIU Xuehan	秦钢强 QIN Gangqiang	王路 WANG Lu	邢雨辰 XING Yuchen	于瀚清 YU Hanqing	赵济宁 ZHAO Jining	朱孟阳 ZHU Mengyang

白雪 BAI Xue	邓诺怡 DENG Nuoyi	高东阳 GAO Dongyang	孔捷 KONG Jie	廖艺敏 LIAO Yimin	马子昂 MA Ziang	孙珂 SUN Ke	王奕绮 WANG Yiqi	杨晟铨 YANG Shengquan	张璐 ZHANG Lu	周金雨 ZHOU Jinyu
曹超 CAO Chao	邓泽旭 DENG Zexu	顾渫非 GU Xiefei	李瑾 LI Jin	林思琪 LIN Qing	麦思琪 MAI Siqi	孙志伟 SUN Zhiwei	王玥 WANG Yue	杨帆 YANG Fan	张鹏 ZHANG Peng	周盟珊 ZHOU Mengshan
程科懿 CHENG Keyi	董冰涛 DONG Bingtao	郭浩哲 GUO Haozhe	李静娴 LI Jingxian	林文倬 LIN Wenzhuo	彭昊韦 PENG Haowei	王奔 WANG Ben	魏高翔 WEI Gaoxiang	杨佳锟 YANG Jiakun	赵茂繁 ZHAO Maofan	祝诗雅 ZHU Shiya
陈露茜 CHEN Luxi	董志昀 DONG Zhiyun	郭珊 GUO Shan	李静毓 LI Jingyu	刘鑫睿 LIU Xinrui	彭洋 PENG Yang	王春磊 WANG Chunlei	谢宇航 XIE Yuhang	杨茸佳 YANG Rongjia	赵文 ZHAO Wen	
陈威霖 CHEN Weilin	范玉斌 FAN Yubin	郝昕苑 HAO Xinyuan	李璐 LI Lu	刘玥蓉 LIU Yuerong	邱国强 QIU Guoqiang	王鲁 WANG Lu	徐小越 XU Xiaoyue	杨乙彬 YANG Yibin	赵晓雪 ZHAO Xiaoxue	
陈宜旻 CHEN Yimin	范悦 FAN Yue	和胤 HE Xu	李帅 LI Shuai	龙千慧 LONG Qianhui	仇凯莹 QIU Kaiying	王琪泓 WANG Qihong	许雁庭 XU Yanting	杨子征 YANG Zizheng	赵越 ZHAO Yue	
陈卓 CHEN Zhuo	方雨 FANG Yu	胡峻语 HU Junyu	李伟 LI Wei	卢卓成 LU Zhuocheng	邱向楠 QIU Xiangnan	王瑞明 WANG Ruiming	徐一凡 XU Yifan	姚孟君 YAO Mengjun	钟言 ZHONG Yan	
翟璺钰 ZHAI Zhaoyu	冯子恺 FENG Zikai	黄柯 HUANG Ke	李文秀 LI Wenxiu	骆靖雯 LUO Jingwen	邵鑫露 SHAO Xinlu	王潇 WANG Xiao	晏攀 YAN Pan	叶庆锋 YE Qingfeng	钟子超 ZHONG Zichao	
崔泂瑄 CUI Yixuan	傅峻岩 FU Junyan	蒋东霖 JIANG Donglin	李宇翔 LI Yuxiang	罗婷 LUO Ting	孙弘睿 SUN Hongrui	王晓茜 WANG Xiaoqian	闫朝新 YAN Chaoxin	章超 ZHANG Chao	周航 ZHOU Hang	

陈晨 CHEN Chen	崔葆莉 CUI Baoli	何家慧 HE Jiahui	孔舒影 KONG Shuying	梁钟琪 LIANG Zhongqi	陆麒竹 LU Qizhu	毛寅初 MAO Yinchu	沈逸哲 SHEN Yizhe	田靖 TIAN Jing	王小元 WANG Xiaoyuan	徐嘉曼 XU Jiaman	喻姝凡 YU Shufan	张雪楠 ZHANG Xuenan
陈浩 CHEN Hao	高世博 GAO Shibo	贺子琦 HE Ziqi	雷雨彤 LEI Yuzhou	林济武 LIN Jiwu	陆禹名 LU Yuming	缪政儒 MIAO Zhengru	宋嘉琦 SONG Jiaqi	田舒琳 TIAN Shulin	汪晶 WANG Jing	杨朵 YANG Duo	岳昭阳 YUE Zhaoyang	赵相宁 ZHAO Xiangning
陈锐娇 CHEN Ruijiao	高爽 GAO Shuang	洪倩倩 HONG Qianqian	李晨 LI Chen	刘传 LIU Chuan	吕强 LU Qiantong	倪钰翔 NI Yuxiang	宋珊珊 SONG Shanshan	汪晶 WANG Jing	文啸 WEN Xiao	杨金铭 YANG Jinpei	詹岳军 ZHAN Yuejun	赵潇艺 ZHAO Xiaoyi
陈睿绮 CHEN Ruiqi	高云剑 GAO Yunjian	胡瑞麒 HU Ruiqi	李富刚 LI Fugang	刘佳慧 LIU Jiahui	罗靖 LUO Jing	潘艺灵 PAN Yiling	孙浩 SUN Hao	王冠一 WANG Guanyi	吴桐羽 WU Tongyu	杨俊岭 YANG Junling	张百慧 ZHANG Baihui	朱秋燕 ZHU Qiuyan
陈雯 CHEN Wen	顾祥姝 GU Xiangshu	胡珊珊 HU Shanshan	李嘉诚 LI Jiacheng	刘瑞翔 LIU Ruixiang	罗宇豪 LUO Yuhao	潘芸娜 PAN Yunna	孙鹤 SUN He	王含宁 WANG Hanning	吴肖行 WU Xiaoxing	杨郡璐 YANG Junlu	张倡铭 ZHANG Changming	朱盈睿 ZHU Yingrui
陈哲 CHEN Zhe	郭辉 GUO Hui	黄佳怡 HUANG Jiayi	李翔 LI Xiang	刘未然 LIU Weiran	吕铁彬 LU Tiebin	彭屹珠 PENG Yizhu	孙穆群 SUN Muqun	王姬 WANG Ji	奚钰竹 XI Yuzhu	杨兆祺 YANG Zhaoqi	张桂煜 ZHANG Guiyu	左琪 ZUO Qi
程琳淼 CHENG Linmiao	郭士博 GUO Shibo	江明慧 JIANG Minghui	李亚斌 LI Yabin	刘文博睿 LIU Wenruibo	马嘉衡 MA Jiacheng	彭幼妹 PENG Youmei	孙强 SUN Qiang	王吕昕 WANG Lüxin	肖雯娟 XIAO Wenjuan	叶芊蔚 YE Qianwei	张佳芸 ZHANG Jiayun	
程意 CHENG Yi	郭烁 GUO Shuo	江天一 JIANG Tianyi	李逸凡 LI Yifan	刘馨琳 LIU Xinlin	马文君 MA Wenjun	沈洁 SHEN Jie	孙欣然 SUN Xinran	王琪琪 WANG Qiqi	谢沙桐 XIE Shatong	游佑莹 YOU Youying	张金库 ZHANG Jinku	
初晓畅 CHU Xiaochang	杭航 HANG Hang	姜辰达 JIANG Chenda	李雨茜 LI Yuxi	刘宗美 LIU Zongmei	毛明权 MAO Mingquan	沈乾诚 SHEN Qiancheng	孙悦添 SUN Yuetian	王思虹 WANG Sihong	徐福锁 XU Fusuo	虞伟炜 YU Weiwei	张新雨 ZHANG Xinyu	

图书在版编目（CIP）数据

南京大学建筑与城市规划学院建筑系教学年鉴. 2022—2023 / 唐莲，李鑫编. -- 南京：东南大学出版社，2025.4. -- ISBN 978-7-5766-1834-1

Ⅰ.TU-42

中国国家版本馆CIP数据核字第202400857D号

编　委　会：丁沃沃　赵　辰　吉国华　周　凌　鲁安东　华晓宁　唐　莲　李　鑫
版面制作：唐　莲　李　鑫　李静怡　沈至文　李若松　黄辰逸　高　晴
责任编辑：魏晓平　姜　来
责任校对：张万莹
封面设计：黄辰逸
责任印制：周荣虎

南京大学建筑与城市规划学院建筑系教学年鉴 2022—2023
Nanjing Daxue Jianzhu Yu Chengshi Guihua Xueyuan Jianzhuxi Jiaoxue Nianjian 2022-2023

出版发行：东南大学出版社
出　版　人：白云飞
社　　　址：南京市四牌楼2号
网　　　址：http://www.seupress.com
邮　　　箱：press@seupress.com
邮　　　编：210096
电　　　话：025-83793330
经　　　销：全国各地新华书店
印　　　刷：南京新世纪联盟印务有限公司
开　　　本：889 mm×1 194 mm　1/20
印　　　张：12
字　　　数：590千
版　　　次：2025年4月第1版
印　　　次：2025年4月第1次印刷
书　　　号：ISBN 978-7-5766-1834-1
定　　　价：78.00元

本社图书若有印装质量问题，请直接与营销部联系。电话：025-83791830